JOHN GLOAG ON INDUSTRIAL DESIGN

Volume 4

PLASTICS AND
INDUSTRIAL DESIGN

T0271086

PLASTICS AND INDUSTRIAL DESIGN

JOHN GLOAG

Taylor & Francis Group

LONDON AND NEW YORK

First published in 1945 by George Allen & Unwin Ltd.

This edition first published in 2023
by Routledge
4 Park Square, Milton Park, Abingdon, Oxon OX14 4RN

and by Routledge
605 Third Avenue, New York, NY 10158

Routledge is an imprint of the Taylor & Francis Group, an informa business

British Library Cataloguing in Publication Data
A catalogue record for this book is available from the British Library

ISBN: 978-1-032-36309-7 (Set)
ISBN: 978-1-032-36590-9 (Volume 4) (hbk)
ISBN: 978-1-032-36609-8 (Volume 4) (pbk)
ISBN: 978-1-003-33287-9 (Volume 4) (ebk)

DOI: 10.1201/9781003332879

Publisher's Note
The publisher has gone to great lengths to ensure the quality of this reprint but points out that some imperfections in the original copies may be apparent.

Disclaimer
The publisher has made every effort to trace copyright holders and would welcome correspondence from those they have been unable to trace.

PLASTICS

and

INDUSTRIAL DESIGN

by

JOHN GLOAG

With a Section on the
Different Types of Plastics,
their properties and uses,
by
GRACE LOVAT FRASER

London
GEORGE ALLEN & UNWIN LTD

FIRST PUBLISHED IN 1945

DEDICATED TO THE MEMORY OF
A GREAT ENGLISH INVENTOR,
ALEXANDER PARKES,
BORN 1813, DIED 1890,
WHO FOUNDED THE
PLASTICS INDUSTRY

PRINTED IN GREAT BRITAIN
in 11-pt. Baskerville Type
BY W. S. COWELL, LIMITED
IPSWICH AND LONDON

Contents

List of Plates

7

Illustrations in the Text

Acknowledgments

SECTION I

For permission to reproduce illustrations in this section, we are indebted to the courtesy of Aga Heat Limited, Allied Ironfounders Limited, E. K. Cole Radio Ltd., De La Rue Plastics Limited, H.M.V. Household Appliances, Halex Limited, Pilot Radio Ltd., Sunbeam-Talbot Ltd. Thanks are also due to Mr. C. F. Merriam for permission to quote some extracts from his broadcast talk, "Plastics—Present and Future."

SECTIONS II AND III

For permission to reproduce illustrations, and for help in collecting material and information, we must acknowledge the courtesy of the American Cyanamid Company (Plastics Division), the *Architectural Forum of New York*, Bakelite Limited, Baker Oil Tools Inc., British Celanese Limited, British Industrial Plastics Ltd., British Resin Products Limited, B.X. Plastics Limited, the Catalin Corporation, the Celanese Celluloid Corporation, the Detroit Macoid Corporation, the Dow Chemical Company, E. I. du Pont de Nemours & Company Incorporated, Durez Plastics & Chemicals Inc., Freydburg Bros.-Strauss Incorporated, the

General Plastics Corporation, Halex Limited, Hercules Powder Company Incorporated (Cellulose Products Department), I.C.I. (Plastics) Limited, Dorothy Liebes, San Francisco, Lumitile, Cincinnati, *Modern Plastics Incorporated*, New York City, Monsanto Chemical Company, Timothy Pflueger, San Francisco, Pierce Plastics Incorporated, Morris Sanders, Schwab & Frank Inc., the Tennessee Eastman Corporation, the United States Department of Agriculture, Bureau of Agricultural and Industrial Chemistry.

DIRECTIONS FOR READING

THOSE who want a compact guide to the various types of plastics should turn at once to Section II, where their properties and uses are set forth by Mrs. Lovat Fraser. Those who wish to examine the possible and actual effects of plastics upon industrial design and commercial art, and desire to know something of the development and to apprehend the significance of the plastics industry, should read Section I, which is written in non-technical language and contains no mathematical or chemical formulæ. Those who dislike reading should ignore Sections I and II and begin at Section III, where "every picture tells a story", and the immense potentialities of plastics are easily perceived.

AUTHOR'S NOTE:

A preliminary outline of the subject covered by Section I appeared in a paper on *The Influence of Plastics on Design*, which I gave before the Royal Society of Arts on May 26th, 1943. The substance of that paper is incorporated in some of my chapters. Some relevant paragraphs from another paper, read to the Design and Industries Association on February 2nd, 1944, on *The Selling Power of Good Industrial Design*, are included in Chapters IV and VI.—J.G.

SECTION I

The Influence of Plastics on Design

By JOHN GLOAG

"It must always be the aim of an industrial organisation to devise and set going one of those systems of manufacture on a large scale with which we have become familiar in recent years. With the aid of suitably designed machinery and methods, great numbers or quantities of some articles in general demand can be produced at a comparatively small running cost. Generally, however, the initial cost is heavy, for the designing of the machinery and the planning of the methods call for great experience and skill, and they demand much time spent in the acquirement of the necessary knowledge and its utilisation in design."

> *Presidential Address to the British Association, September* 5, 1928, *on* Craftsmanship and Science, *by Professor Sir William Bragg.*

"I would suggest that it is quality of design, even more than quality of material, which gives sustained value and interest. Bronze, for instance, is a noble material, but if something made of bronze is poorly designed it is boring instead of stimulating, and is soon neglected, or used merely as a convenience and only until something better is produced. The same principle applies to plastics. It is qualities of style and of design which confer reasonably sustained satisfaction."

> *Contribution to a discussion on* The Influence of Plastics on Design, *by Percy Smith, R.D.I. Journal of the Royal Society of Arts*, Vol. XCI, No. 4644, page 470.

"Milk from plastics!   What next?"

PLASTICS AND INDUSTRIAL DESIGN
SECTION I *by* JOHN GLOAG

CHAPTER I

The Plastics Family : Origin and Character

FEW materials used or made by man have aroused so much general interest or so many speculations regarding their future uses as those chemically-produced substances, capable of being shaped by the application of heat and pressure, which are now known as plastics. Their diversity is bewildering; their capabilities both impressive and stimulating; and many people are prepared to believe that they can do practically anything, a state of mind that has been neatly satirised by *Punch* in a picture which shows an innocent gentleman raising a brimming glass in front of a complicated machine and exclaiming: "Milk from plastics! What next ?"

It is the purpose of this book briefly to examine plastics in relation to industrial design. I should explain that I have neither the qualifications nor the knowledge to write with technical authority on the subject of plastics. Although I touch on the past history and future prospects of the plastics industry, I am in no sense a technician, and my section of this book is chiefly concerned with the effect of plastics upon contemporary industrial design, and in particular with methods whereby the use of these materials may be jointly explored in the future by manufacturers and industrial designers. The various types of plastics, their characteristics, properties, uses and methods used for fabricating them are reviewed by Mrs. Lovat Fraser in Section II; while Section III illustrates applications and possibilities drawn from British and American sources.

After the first World War the rayon industry was established: its influence has been universal. During the second World War the plastics industry has attained a comparable importance. There is hardly any large-scale industry in Britain and the United States which does not now use plastics in some form. So enthusiasts who announce on every appropriate and inappropriate occasion that we are about to enter "the age of plastics" are powerfully supported by facts, though their announcements would have a keener flavour of realism if they added the words: "and light alloys." The new materials that have developed during the first forty years of this century are by no means covered by the word "plastics." Aluminium and its alloys and magnesium have properties that are as astonishing and revolutionary as those possessed by the plastics family; and older materials, such as glass, timber—particularly in the form of plywood—cast iron and steel, have acquired new characteristics and properties as the result of an immense programme of creative research work that material-producing industries have initiated.

The modern plastics industry is of comparatively recent growth. The experimental work that led to its inception and ultimate development began in Birmingham in the middle of the nineteenth century, when the first patent bearing on "celluloid" was taken out in 1855 by Alexander Parkes. "Celluloid" is the oldest plastic, and it was developed commercially during the late nineteenth century; but the event which marks the beginning of the plastics industry as we know it to-day, was the patent taken out in 1909 by Dr. Leo Hendrik Baekeland for the plastic known as "Bakelite." Dr. Baekeland was born in Ghent in 1863. Two years later, Mr. Alexander Parkes, of Birmingham, gave to the Society of Arts what was probably the earliest description of the possibilities of plastics. On December 20th, 1865, he read a paper to the Society on "The Properties of 'Parkesine,' and its application to the Arts and Manufactures." In the opening paragraphs of that paper, he said: "For more than twenty years the author entertained the idea that a new material might be introduced into the arts and manufactures, and in fact was much required; he succeeded in producing a substance partaking in a large degree of the properties of ivory, tortoise-shell, horn, hardwood, india-rubber, gutta percha, etc., and which will, he believes, to a considerable extent, replace such materials, being capable of being worked with the same facility as metals and wood. This material was first introduced, under the name of Parkesine (so called after its inventor), in the Exhibition of 1862, in its rough state, and manufactured into a variety of articles in general use; it then excited the greatest attention, and received a prize medal . . . ."

Mr. Parkes explained that Parkesine was made from "pyroxyline and oil, alone or in combination with other substances." The various degrees of "hardness or flexibility" were "obtained in the easiest and most expeditious manner by varying the proportions of pyroxyline, oil, and other ingredients."

In that interesting and highly technical paper, Mr. Parkes described what we can now recognise as a forerunner of the materials that are known to-day under the generic name of plastics. He described to his audience the difficulties and discouragement he had encountered. He said: "The innumerable trials and investigations required, involved no less than twelve years' labour and an expenditure of many thousand pounds, before the material could be proved to be really of commercial importance; and although this may appear a long time to pursue one object, the author wishes to explain that time has been itself of the utmost importance in developing this manufacture, as it has enabled him to test the effect of time on the material, and also of atmospheric changes, and many other influences; this has proved of great value in arriving at his present knowledge of the material."

He ultimately produced Parkesine, in some of its qualities, at 1/- per lb. He gave his hearers a glimpse of the "age of plastics" when he said: "The applications of this material to manufactures appear almost unlimited, for it will be available for spinners' rolls and bosses, for pressing rolls in dyeing and printing works, embossing rolls, knife handles, combs, brush backs, shoe soles, floor-cloth, whips, walking sticks, umbrella and parasol handles, buttons, brooches, buckles, pierced and inlaid work, book-binding, tubes, chemical taps and pipes, photographic baths, battery cells, philosophical instruments, waterproof fabrics, sheets, and other articles for surgical purposes, and for works of art in general.

"There is one application of Parkesine which, as far as experiments have gone, promises to be of great importance, viz., insulating telegraph wires. It will be at once evident, from the nature of the ingredients used, that by simple mechanical and chemical processes, perfect freedom from impurities or foreign ingredients can be attained; a most important property in a material which it is intended to employ for electrical and insulating purposes. . . ."

Those who wish to refer to his paper will find it in the *Journal of the Society of Arts*, Vol. XIV, No. 683, December 23rd, 1865, which also contains a record of the subsequent discussion, during which a Dr. Bachhoffner protested "against inventions of this kind being called after the name of the inventor." The Chairman, Mr. William Hawes, in proposing a vote of thanks, took occasion to say that "so far as his own individual opinion went, he thought it desirable that papers read before a Society like this should not have for their object merely the description of some particular invention, but should be more general in their character. No one, however, could question the importance of a discovery which introduced a material likely to be of great value in the arts and manufactures of the country, and the want of which was becoming more and more felt. We were exhausting the supplies of india-rubber and gutta percha, the demand for which was unlimited, but the supply not so. In the case of gutta percha the tree was destroyed in taking the produce of it, and we had to wait till other trees grew for future supply; and with regard to india-rubber, the plants only produced a limited quantity each year. This new commodity, however, was produced from materials of which there was an unlimited supply. . . ."

Alexander Parkes was one of those men of genius who not only possess original creative powers, but are endowed with the hard common sense which urges them to apply their gifts to the solving of practical problems. He was the son of a lock-maker; and the astonishing fertility of his mind found expression in many inventions. Before he died in 1890, he had taken out some eighty patents, covering an enormous range of activities, and including metallurgical subjects, electro-plating, nitro-cellulose and "celluloid", furnaces, candles and the ornamentation of metals. In some ways he resembled those adventurous and speculative scientists who, in the latter half of the seventeenth century, formed a little club which ultimately became the Royal Society.

When he lectured a few days before Christmas in 1865 to the Society of Arts, nobody could have foretold that the substance called Parkesine was the progenitor of the most versatile family of materials that has ever served the needs of mankind. The Chairman of that meeting of the Society would certainly not have recognised the word "plastics"; but just over fifty years later the Perkin Medal for 1916 was awarded to Dr. Leo Hendrik Baekeland, B.S. and Sc.D., "for his distinguished services in the fields of photography, electro-chemistry and plastics" at the regular meeting of the New York Section of the Society of Chemical Industry.* In that year, "the first moulding powder was formed which heralded the advent of the new plastics industry."† But the word plastics was not yet common currency; and it was only during the nineteen-thirties that it became widely known in Great Britain, although plastics had been fabricated in the country for over seventy years. They

* January 21st, 1916. The proceedings of the meeting, and the Medal Address given by Dr. Baekeland are recorded in full in *The Journal of Industrial and Engineering Chemistry*, Vol. VIII, pages 177-179. (Published by The American Chemical Society).

† *Plastics*, by Dr. V. E. Yarsley and E. G. Couzens. Chap. I, Section IV, page 16. (Pelican Books, 1941).

have now reached a stage of development when "they form a fifth class to the materials, metal, wood, glass and ceramics used in the past." That statement was made by the President of the Royal Society of Arts, Dr. E. Frankland Armstrong, in a paper entitled *Materials Old and New*, read before the Society in January, 1942.

Between 1935 and the beginning of the second World War, considerable attention was given to plastics, both in the national press and in many technical and professional journals, though the true character and capacity of the various types were not always fully apprehended by those who spoke of them, with or without authority. In this country—indeed in any country inhabited by human beings—a new idea has to adopt some form of protective mimicry if it is to survive the frosty reception generally accorded to any suggestion of change. For some years plastics masqueraded as wood, or metal, or leather, or glass. They were introduced in the guise of substitutes, for although industry in the past has often feared and rejected originality, it has welcomed imitation. Mr. Alexander Parkes, when he broke the news about "Parkesine," tactfully likened it to such familiar materials as ivory, tortoise-shell, horn, hardwood, india-rubber, and so forth. Until 1935 only comparatively few people had recognised plastics as a new and independent group of materials, and even after that date some nervous manufacturers persisted in making them do their worst, which is to imitate traditional materials, instead of their best, which is to bring to the service of industry an array of new properties—new gifts of lightness, translucency, transparency, texture and colour.

I have mentioned 1935 as the beginning of a period when the British public became actively conscious of the existence of the plastics family. During that year an exhibition of British Art in Industry was promoted by the Royal Society of Arts in collaboration with the Royal Academy, and was held in London at Burlington House. A spacious section was devoted to plastics, and the display of that section was designed ably and vividly by Mr. Grey Wornum, F.R.I.B.A. Among the various plastics assembled in that room was one exhibit that aroused particular interest. It had an unusually dramatic quality: it suggested remarkable possibilities. It was a piece of colourless material—an irregular lump of pure transparency. This exhibit, with its odd depth of transparency, was labelled "Resin M." It was described as a synthetic resin. It reminded you rather of still water in a clear and shallow pool. Although it resembled glass in its transparency and freedom from colour, it lacked the flash and sparkle of glass.

It should be said here that most transparent plastics have a slightly lower refractive index than that of glass, though this is not the chief reason for the difference in degree of surface brightness. The main difference lies in the quality of the surface. When a conchoidal fracture is made in a piece of glass, the edges and surface irregularities of the glass are always brilliant, while a similar fracture in a piece of transparent plastic, such as methyl methacrylate,* is duller both at the surface and the edges, and becomes progressively less brilliant with time. The surface of a moulded plastic takes on every detail of the mould, and to produce a bright surface the mould must itself be finished to a high polish. But sooner or later the surface of the moulding becomes dulled, sometimes due to shrinkage and sometimes due to slight absorption of water, either of which will cause minute surface flaws or irregularities resulting in loss of brilliance: these effects are quite apart from the slight roughening which almost inevitably occurs during ordinary handling and cleaning. Sheets of certain of the transparent plastics are more transparent than glass because they allow a higher transmission of light, particularly at the two ends of the spectrum and noticeably in the region of the blue. The greater sparkle of glass as seen by reflected light is due to the higher refractive index and consequent higher surface reflection of glass, the effect being particularly emphasised when there is any bend or unevenness of surface.

The significance of the refractive index is lucidly expounded in *The Invisible Man*, when H. G. Wells makes Griffin, the unhappy hero of that fantasy, describe to Kemp the relative visibility of various organic and inorganic materials. Griffin says: "You know quite well that either a body absorbs light or it reflects or refracts it or does all these things. . . . A diamond box would neither absorb much of the light nor reflect much from the general surface, but just here and there where the surfaces are favourable, the light would be reflected and refracted so that you would get a brilliant appearance of flashing reflections and translucencies. A sort of skeleton of light. A glass box would not be so brilliant, not so clearly visible as a diamond box, because there would be less refraction and reflection. See that? From certain points of view you would see quite clearly through it. Some kinds of glass would be more visible than others—a box of flint glass would be brighter than a box of ordinary window glass. A box of very thin common glass would be hard to see in a bad light, because it would absorb hardly any light and refract and reflect very little. And if you

* See Section II, page 78.

put a sheet of common white glass in water, still more if you put it in some denser liquid than water, it would vanish almost altogether, because light passing from water to glass is only slightly refracted or reflected, or, indeed, affected in any way."*

A transparent plastic would have furnished Griffin with another example, to illustrate the progressive lowering of the refractive index, before he reached the stage of submerging a sheet of glass in water.

Transparency has an almost universal appeal; and the unusual nature of that transparent material in the plastics section of the Burlington House Exhibition was the starting point of innumerable trains of thought, particularly among imaginative designers. It also helped to register a rather misleading impression, for many people thereafter thought of plastics chiefly as transparent materials. Another impression made by the Exhibition was the ease with which plastics could be moulded. This impression was certainly not misleading; by compression moulding or injection moulding, according to the type of plastics, an infinity of shapes are available. The fabrication of plastics depends partly upon the form of the raw materials, and the basic forms are powder, sheet, tube or rod.

A brief description of the main groups of plastics is desirable here, if the effect of these materials upon industrial design is to be studied.

Plastics may be divided into two main groups: *Thermo-plastic* and *Thermo-setting*. There is a third group, based on the Protein Caseins.

All plastics are capable of being moulded under heat and pressure. When *Thermo-plastic* materials are thus moulded they do not change chemically, and may be reheated and reformed. This means that scrap can be used and consequently waste may be eliminated. *Thermo-setting* plastics, when once they have been shaped by heat and pressure, cannot be changed; they remain hard, and can only be worked by cutting or grinding.

Now each of those two main groups—thermo-plastic and thermo-setting—includes various types which possess the fundamental character of their group, but differ from each other both in quality and in the sources from which their raw materials are drawn. The thermo-setting group includes phenolic resins, cast phenolic and urea resins. In the thermo-plastic group are cellulose nitrate, cellulose acetate, acrylic resins and polystyrene. The protein plastic group is derived from casein.

I do not propose to deal with raw materials or methods of manufacture, or to anticipate Mrs. Lovat Fraser's review which follows Section I, but to indicate the range of plastics, I will outline

* *The Invisible Man*, by H. G. Wells, Chap. XIX.

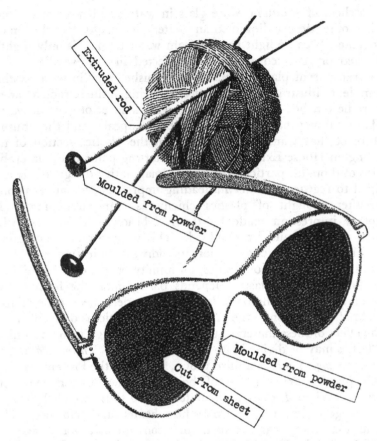

The basic forms in which plastics are available for working are: powder, sheet, tube or rod. The sheet may be either a solid slab or a film, 3/1000ths of an inch thick; the rod may be a hair-fine filament. Here are some common objects cut from sheet, moulded from powder and made from extruded rod.

briefly the main characteristics of those covered by the three groups.

In the *Thermo-setting* group are the following:—*Phenolic resins*: They have resistance to shock, heat, acid and alkali. They are free flowing for extrusion or transfer moulding. Their chief use is in compression moulding.

Cast phenolic: This has high impact strength, rigidity, is non-inflammable, and is easily coloured and fabricated.

Urea resins: They are translucent and have a large colour range; they are splinter-proof, light in weight and rigid. They have a hard surface and they diffuse light well. They are almost without taste or smell.

The *Thermo-plastic* group includes the following:—*Cellulose nitrate*: It is tough, but easy to fabricate, it is also easy to cement and to colour. It resists the action of water and is transparent. *Cellulose acetate*: It is tough, light, alkali-resisting, odourless, and can be easily fabricated and coloured. *Acrylic resins*: They have rigidity and lightness, they resist water and weather, and take colour easily. They have exceptional transparency. *Polystyrene*: This gives a great range of colours, and can be either transparent or opaque. It has a high index of refraction; it can transmit light, and it resists chemical action.

Finally, there is the *Protein-plastic* group, based on casein. This group includes plastics that are non-inflammable, easy to colour, polish and fabricate.

So much for the origin and general character of the plastics family. In Section II, Mrs. Lovat Fraser records in detail the characteristics of the individuals in the different groups of that family.

CHAPTER II

Plastics are not Substitutes

EVEN the highly condensed account of the growth of the plastics industry and the general character of the plastics family given in the first chapter suggests the perplexing possibilities provided by these synthetic materials. Apart from our normal, human resistance to unusual things, the inclination to regard plastics chiefly as substitutes is often supported by the hope that the rich confusion of the prospect they disclose may thus be simplified. The most august technical authorities sometimes display a tendency to emphasise the traditional materials that plastics could displace, even when they are elucidating the scope and novelty of their inherent properties. For example, Professor J. D. Bernal, in discussing the gifts that the chemical industry may bestow upon the world, writes: "Already we have, in artificial silk and plastics such as bakelite, materials which are successfully displacing natural products; and this tendency may become more general."*

The accent here is on *displacement*; and although scientists are notoriously cautious, displacement is the main theme of the last chapter of one of the most informative books yet published on this subject. Dr. V. E. Yarsley and Mr. E. G. Couzens, the joint authors

* *The Social Function of Science*, by J. D. Bernal, F.R.S. (George Routledge & Sons, Ltd. third printing, 1942). Chapter XIV, "Science in the Service of Man," page 371.

of *Plastics*,* have obviously enjoyed themselves enormously by making
a clean sweep of old materials when they discuss "Plastics and the
Future."† They imagine the dweller in the "Plastic Age" coming
"into a world of colour and bright shining surfaces" and conduct
him from the smooth, sleek serenities of his plastic nursery through
his three score years and ten, when he is neatly tucked up, or, as
the authors put it, "hygienically enclosed" in a plastic coffin.‡ Like
Mr. Mandragon, the Millionaire in G. K. Chesterton's poem, his
death will be as sheltered and artificial as his life; and he will
lie in the grave, "hygienically enclosed" and protected and

> ". . . certainly quite refined,
> "When he might have rotted to flowers and fruit with
>   Adam and all mankind . . ."

It is undesirable that one class of materials should ever dominate
our environment: however versatile the components of that class
might be, their universality would destroy variety, and, in the words
of Mr. Winston Churchill ". . . the most fertile means from which
happiness may be derived in life is from variety."

It is clear that in some directions plastics will replace natural
products. For instance, nylon§ bristles have already replaced hogs'
bristle in tooth-brushes and hair-brushes. We are not always aware
of the extent of the service plastics are already performing in everyday life nor how far they influence or control the technique of
industrial production. An excellent summary of the subject was
furnished by Mr. C. F. Merriam in a talk entitled "Plastics—
present and future," broadcast in the B.B.C. Home Service on
March 13th, 1942. Mr. Merriam, who is an industrialist concerned
with the fabrication and production of plastics, asked his listeners
to imagine what would happen if there was a sudden and miraculous
withdrawal of those materials from the world. He said: "If you
were using a modern telephone at that moment, the whole instrument, with the exception of a few small metal parts, would disappear.
You wouldn't be able to go on knitting that sock you're holding,
because the plastic needles would have vanished. If you were
playing a game of table tennis, you'd find yourself hitting the air
instead of the ball. If we were eating dinner, many of us who
happen to wear false teeth would find it most uncomfortable because
the dentures would have gone, leaving the teeth loose in the mouth.
If you were riding a cycle or driving a car, the handle-grips or

* *Plastics* (Pelican Books, 1941), quoted on page 16.
† *Ibid*, Chapter IX, pages 151-158.
‡ *Ibid*, page 158.
§ See Section II, page 93.

steering wheel would suddenly feel cold and harsh because the plastic covering had been removed. The wireless set you're listening to might fade out because the case and some vital part had gone. The comb you might be passing through your hair would also vanish. And what about those buttons on your dress and those coloured zip-fasteners that hold you together? They would go as well . . . Perhaps the most astonishing result of this miracle would be that air warfare would cease for a time because pilots and crews of very fast aeroplanes couldn't withstand the rush of air, from which they are now protected by plastic windows and cockpit covers. And so you see plastics enter into our daily lives in quite a number of unexpected ways. I could give you many more instances, but I think I've said enough to show you that without plastics life would be rather different."

Professor Bernal tells us that the chemist, as a result of "the advance in theoretical chemistry, particularly in structural and colloidal chemistry," may be able "to plan the structures of materials according to desired properties";* Mr. Merriam in the talk just quoted, described the development of one of these new, chemically produced materials. "New plastics," he said, "are generally discovered or developed by the concentrated effort of a number of men over a long period. The most striking example of this . . . is the work of an American chemist who, with more than two hundred other chemists and engineers, was engaged on the research which produced a plastic called nylon. It took them about ten years. Nylon is used in America for women's stockings; they are like silk but quite a different material from rayon; they are strong and less likely to ladder . . ."

Glancing at the future, Mr. Merriam mentioned an experimental motor-car that had been made, with the body-work wholly of plastics mounted on a tubular frame. "It is as strong as the present-day metal body and much lighter, so that a less powerful engine is sufficient. Another advantage, of course, is that, being self-coloured, a plastic body does not need to be painted . . . Small boats of plastic have been made, though to what extent plastics will enter into the construction of larger vessels I wouldn't like to say, except for the fittings and decorations, of which, for instance, there are £20,000 worth in the *Queen Mary*."

Another restrained examination of possible future applications of plastics, with particular reference to the building industry, appeared in a survey published by the *Architect's Journal* on October

*The Social Function of Science, Chapter XIV, page 371.

29th, 1942. It was suggested that large, finished components might be provided by such materials as cellulose acetate* and the styrene resins.† The significance of this application is conveyed by the following quotation from this survey: "Units of building equipment, of a comparatively large size, have been made by quite a number of methods. Whilst thermo-setting resins lend themselves to compression moulding processes, the thermo-plastics can be produced quickly and efficiently by *injection moulding*, a process which makes it possible to carry out the most intricate designs in a single operation."

The architect certainly contemplates the extensive use of plastics in building, and the large, finished components referred to suggest that plastics may play a not inconsiderable part in the production of pre-fabricated building units. In a discussion entitled "New Buildings for a New Age," broadcast in June 1941 between two architects, Christian Barman and Grey Wornum, and myself, with Joseph MacLeod as the compére, Mr. Wornum said:

"As an architect, my interest in plastics lies in the new constructional products for which the future will offer the most tremendous possibilities. For example: it will be perfectly possible to have bathrooms with not only the small fittings, but many of the large ones, including the plumbing pipes, created of plastics. But I must say something about the manufacture of the material—I promised not to be technical—so that you can get a better idea of how it can be fabricated. You see, the steel presses for these moulds are very costly, and that means that you have got to have mass production in order to justify the initial cost of the processes. Now, if you are going to mass produce anything with a new and wonderfully fluid material, you must design it decently in the first place, and you should avoid imitating other materials.

"Plastics are often made to look like wood or marble, in spite of being their own good-looking selves. If we make the most of the character and the finish of the material, then we have got something completely new with which to design all kinds of interior fittings in a house."

He also mentioned that "Electric heating wires have been successfully embedded into laminated sheets to prevent ice forming on aeroplane wings, and this may well develop in post-war days for heating panels for rooms and cupboards, even for heated bed ends. If large aeroplane parts can be so produced, it will be but

* See Section II, page 73.
† See Section II, page 80.

a step farther to make in plastics whole sections of our houses as mass-produced units."*

In another book I have discussed the effect of different types of new materials on design, and I quote the following from a relevant chapter:

"The vast interest in the potentialities of plastics that was acquired by responsible people in British industry in the years that immediately preceded the second world war, had a profound and beneficial effect upon the British war effort . . . The war accelerated the development of the industry. Hundreds of new uses were found; research and experiment were encouraged; thousands of problems connected with aircraft production were urgently needing rapid solution; supplies of most normal and accepted materials were diminished or interrupted, or could not meet the expanded demands of war-time industry. So plastics came into the picture—first (as in peace-time industry) chiefly as substitutes for familiar materials, and then, as invention and development swept along, they became materials in their own right. Achievements that would have been considered impossible in 1939, methods unthought of when the war began, substances and combinations of substances that were only in the theoretical stage ten years ago, became, by the third year of the war, almost commonplace.

"The research chemist and the production engineer went into partnership, as metallurgists and engineers had gone into partnership over a hundred years earlier. That century-old association of technicians had produced a new world of steel, which altered the technique of shipbuilding and architecture. To-day, the chemist and the production engineer have brought nearer a world of new weights and measures, new ways of withstanding stresses and strains, new resistances to wear, tear and changes of temperature, new tensions. The vigour and development of the plastics industry in Britain, the enterprising and progressive outlook of the men who are directing it, and the inventive fecundity of its technicians, suggest that we may look forward to changes in the familiar shapes of innumerable things. New qualities, such as translucency and transparency, will be conferred upon the objects of everyday life, and on industrial products. Such anticipations are not mere flights of imagination; some hard-working reality of practical achievement already sustains them, for the British plastics industry in war-time has proved that these vastly varied materials can stand up to tasks

* B.B.C. Home Service feature, *Ariel in Wartime*, June 1941. Reproduced in full in *Plastics*, Vol. V, No. 49, June 1941, pages 111-113.

so exacting that normally only hard metals were supposed to be endowed with the endurance demanded for their effective and safe performance. Plastics are used for aircraft windows and cockpit covers, tank windows, aircraft stowages, eyeshields, floating torches, emergency ration packs, map cases; and some of the latest types are replacing rubber for insulation purposes. That is only the beginning of a huge list of uses."*

The windows and cockpit covers of aircraft are made from transparent plastics belonging to the cellulose acetate† and acrylic‡ groups. They are used because of their extreme lightness, and, in the case of the acrylics, because of their splinter-proof qualities and their perfect clarity. This is a practical application of that new transparency, foreshadowed by the exhibits at Burlington House in 1935. These qualities, so vital in their war uses, may give service in the new world fleets of freight and passenger aeroplanes; some windows of trains and motor coaches may eventually gain benefit from them, and new sunshine roofs in cars may derive both form and material from war-time prototypes—from the domed, transparent cockpit covers and gun-turret hoods of fighting aircraft. Before the second world war, the curved windows of the beaver-tail observation cars of the L.N.E.R. "Coronation" Express were made of transparent plastic sheet. The use of plastics in alliance with aluminium and its alloys may have great economic significance for the design of vehicles, because the lightness of such materials increases the pay-load. A cheap motor car may become a glistening, mobile cell; a tiny engine may do fifty or more miles to the gallon, because it is bearing a featherweight four-seater saloon of plastics and aluminium, while tough, long-wearing plastics perform quiet, efficient work in the gear-box.

When an industrial designer uses plastics to the best advantage, he achieves economy of weight and a new smoothness of surface, and if he should employ transparent or translucent materials, the shapes he devises melt into partnership with light, natural or artificial, thus attaining a wholly new decorative quality. The styrene resins§ and the acrylic resins¶ are materials that possess, apart from all their other properties, attributes that are comparable only with such natural substances as amber, jade, ebony and ivory.

* *The Missing Technician in Industrial Production,* by John Gloag. (George Allen & Unwin, Ltd., 1944), Chap. VIII, pages 78-80.
† See Section II, page 73.
‡ See Section II, page 78.
§ See Section II, page 80.
¶ See Section II, page 78.

They do not resemble those substances; but they possess a similar richness of ornamental character. It is this aspect of their character that gives a special significance to Dr. Frankland Armstrong's statement, quoted in the previous chapter, that plastics form a fifth class to the materials, metal, wood, glass and ceramics. New forms appear when creative, experimental research work in design is conducted with this "fifth class" of materials.

To regard them as substitutes hampers their development; to regard them as the only satisfactory answer to every problem of design, industrial, architectural and even sartorial, perverts their use.

CHAPTER III

"Limitless Control of Material"

MANY people appear to think quite seriously that plastics are going to be the only materials worth bothering about in the future. This state of mind, which is not so much a state of mind as a state of intoxication, may lead to a widespread misuse of plastics in industry; it will certainly encourage the growth of mushroom enterprises for the fabrication of the materials, and uncritical investors will pour cash into such concerns, pleasantly fuddled by the belief that anything connected with plastics means easy money. The get-in-and-get-out technique of financial adventurers, who disgrace the name of free enterprise and know nothing of creative industry, may discredit plastics and hamper their progressive development by reputable firms. Such plunderers may do infinite harm to the plastics industry, for they will exploit the credulity and interest of the public, who, after losing their money, may lose their interest, and perhaps some of their credulity. (For some reason or other, these predatory types are identified as representative "business men" by the opponents of private enterprise; though they would hesitate to assert that the jerry builder or the land speculator represented the architectural profession, or that a vendor of harmful or useless patent medicines was a typical physician).

Another danger arising from an intemperate enthusiasm for plastics, is the assumption that their extensive use will automatically turn familiar, traditional materials into back numbers. Some writers have committed themselves to the statement that plastics will largely replace glass, which is about as sensible as saying that trousers will replace coats or chairs will replace sideboards. Sweeping statements of this kind could be made only by people who persist in regarding plastics as substitute materials. Again, such people imply that light alloys will be ousted by plastics; an unbalanced belief, and as basically unreasonable as saying that cheese will replace bread or bacon will replace eggs.

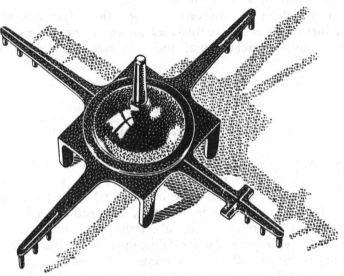

A delicate moulding such as this demonstrates the "limitless control of material" rendered possible by plastics. It is made from cellulose acetate. (*See* Section II, Fabrication Methods, page 104).
(*This drawing, made from a photograph, is reproduced by courtesy of Halex Limited*).

To any student of the technique of industrial design, it is obvious that many materials have complementary uses, and that productive partnerships between materials are not only possible, but almost inevitable. In the course of war production, many such partnerships have been established between old and new materials, some of them resulting from war-time research work. For instance, aluminium has been welded to glass; combinations of plastics and light alloys have been used; plastic sheets and plywood have been cemented together; plywood has also been impregnated with various synthetic resins, so the best of both materials is available. Resin-bonded plywood preserves much of the character of wood, and eliminates many of its disadvantages. These various partnerships, and the changes wrought in the traditional character of old materials, allow the designer not only to create objects that are perfectly adapted to perform specific functions, but *to create the material* from which they are made. To quote again from the survey of plastics in the *Architect's Journal*, this time from an imaginary debate between a pessimist and an optimist on the future of resin-impregnated materials: "With the extensive use of plastics the prospect is nothing less than that we shall be able to 'make' wood into the material we want. The specific weight, the tensile strength, resistance to moisture, electric properties and all other features can be simply designed for the purpose, and the material 'made to measure.' You may not consider such products entirely as plastics, as the proportion of timber or other fibrous material may exceed the resin-content, but the opportunity of making such 'built-up' materials is only possible by using plastics."*

These views are reinforced in a paper by Mr. R. J. Schaffer, of the Building Research Station (Department of Scientific and Industrial Research), read before the Institute of the Plastics Industry on February 17th, 1942.† He refers particularly to resin-bonded plywood as a promising material for prefabricated building units, for it is highly resistant to moisture and may be used externally. He quotes the term "improved wood" in connection with thin wood laminations which are impregnated and subsequently bonded under pressure.

The implications of this new capacity to make materials to measure, and the growing powers of the chemist to create substances with properties selected and adjusted to fit the most exacting specifications, must be appreciated alike by manufacturers and designers; not with the sort of appreciation that begins and ends with an enthusiastic welcome for the manifold conveniences that will ease and stimulate their respective tasks, but with sober under-standing of the chaos that may result from the reckless and unimaginative mood, when people say: "These materials can do *anything*—let's do *everything*!" Ideas that begin in the mind of some inventive genius, like Alexander Parkes, are apt to grow in all sorts of unexpected and embarrassing ways; but it is rare for any scientist to foresee the ultimate impact of his theories and experiments on the habits and beliefs of his fellow men. Great gifts and murderous dangers leak out of the research laboratory; and since the end of the last century only one notably articulate Englishman has been consistently aware of this fact, and his name is H. G. Wells. A pupil

---

* *The Architect's Journal*, Vol. 96, No. 2492, page 286. See Chapter II, page 23.
† *The Use of Plastics in Building*, by R. J. Schaffer. Printed in *Chemistry and Industry*, Vol. LXI, No. 34, pages 357-361. (August 22nd, 1942).

of T. H. Huxley, he has been the prophet, though he has failed to be the guide, of what he has aptly named "the scientific commercial age." For half a century he has been the source of a stimulating torrent of ideas and suggestions about the effect of applied science on social and economic life. With an imagination far superior to Edgar Allan Poe or Jules Verne, with a human understanding comparable with that of Charles Dickens, he has seen how the pattern of everyday life could be changed in the laboratory. In his scientific romances he has often accurately anticipated the effect of some discovery or industrial technique; many of his prophetic ideas have become established facts, some of them unfortunate facts, as anyone to-day may fully appreciate by re-reading *The War in the Air*, which he wrote in 1908. As long ago as 1902 he was writing, in *Anticipations*, of materials that would enable builders to use large-scale prefabricated units for the quick erection of houses. In 1923 in *Men Like Gods*, he described a delectable Utopia, where all the harassing economic problems of our century had been comfortably solved, and everybody enjoyed the material benefits that consistent and unrestricted scientific research may one day bring to mankind. Into this serene and tidy Eden he introduced some earthly visitors; and one of them, contemplating the equipment of his Utopian bedroom, found that "the forms of everything were different, simpler and more graceful." His delighted surprise brought him to these conclusions: "On earth, he reflected, art was largely wit. The artist had a certain limited selection of obdurate materials and certain needs, and his work was a clever reconciliation of the obduracy and the necessity and of the idiosyncracy of the substance to the æsthetic preconceptions of the human mind. How delightful, for example, was the earthly carpenter dealing cleverly with the grain and character of this wood or that. But here the artist had a limitless control of material, and that element of witty adaptation had gone out of his work. His data were the human mind and body. Everything in this little room was unobtrusively but perfectly convenient—and difficult to misuse."

Plastics will give the designer precisely that "limitless control of material." Hitherto it has been attained only by enormous expenditure of time and labour. There have been periods when designers have rebelled against the sobering discipline imposed upon them by the nature of materials. The Rococo phase of French taste in the eighteenth century, the furniture of the Louis-Quinze period, disclosed a febrile appetite for luxury, an almost delirious desire for complexity in form and ornamentation; designers had then, at vast cost, attained a dubious freedom from the limitations of materials; they ignored the fact that limitations existed, and where it was impossible to ignore some stubborn resistance or incapacity in wood or metal, they were ingenious in the methods they invented for coaxing or bullying the recalcitrant material into some predetermined and unsuitable shape. Writing of that period, a great critic, Lisle March Phillipps, has said: "Seriousness in life and art

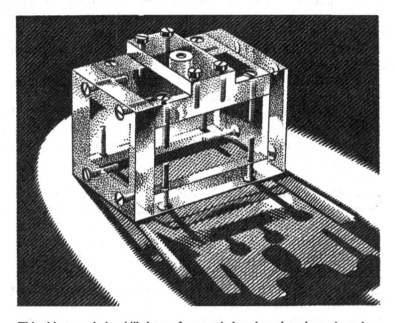

This object, made by skilled men from methyl methacrylate sheet, shows how the individual craftsman may be endowed with fresh powers and opportunities for the exercise of his skill by some plastics. If these materials give the designer a new and almost unbelievable liberty for experiment and expression, the specialist craftsman must inevitably gain a corresponding freedom.
(*This drawing, made from a photograph, is reproduced by courtesy of Halex Limited*).

goes out with Louis-Quatorze; frivolity comes into life and art with Louis-Quinze. The old strength and stateliness gives place to an artificial and excessive refinement in workmanship, not of detail only but of form. What was ornament in the older style assumes control, eats form away, until form itself becomes ornament. It is the peculiarity of the studies of curves and scroll work of Louis-Quinze furniture, and the slender, attenuated proportions of Louis Seize, that they no longer represent the beautifying and perfecting of the common things of life, which, after all, is the true function of art as applied to things like furniture, but minister and bear witness to a life cut off from such things. It is impossible to associate these exquisite creations with the idea of everyday life and common use at all. They have forgotten all about use and reality, and have made of mere luxury their *raison d'etre* and supreme justification. The artificial has to them become the real."*

We know the price paid by that fantastic society for its deliberate disregard of realities. The society disappeared, and its exuberant essays in ornate decoration remain to remind us of what happens to designers and to design when the artificial becomes the real. Incidentally, that forcing and bullying of materials did not always produce things that could endure: in many of those extravagant pieces of Louis-Quinze furniture the veneers and the delicate inlays of precious materials have cracked, and the frothing bronze embellishment has parted company from the woodwork.

However firmly those French designers resolved to ignore the limitations of materials, they could never quite succeed in doing so. The designing and making of furniture has always been conditioned by the nature of the materials available. Wood has to be chosen carefully; it is never static; however ancient it may be, even if hundreds of years have elapsed since the tree was felled and the boards first seasoned, wood will always move, for it absorbs moisture and swells and, with a change of temperature, it will contract as the moisture dries out. The skill and knowledge of woodworkers have always been directed to overcoming, or at least minimising, this inherent characteristic of timber. Joints and framing have been devised to give rigidity and to correct the tendency of large boards to warp; and in this age-old struggle with timber, woodworkers have created a superb technique of cabinet-making and joinery. Such difficulties have always stimulated the craftsman to fresh inventive efforts, and the whole technique of woodworking, which is one of the special gifts of skill England has contributed to civilisation, has been improved, generation after generation, because timber had to be studied and cajoled into the service of man. Iron, lead, the alloys of copper—bronze and brass—all challenge the ingenuity of the craftsman; they have to be driven by hard work and exacting technique to do their job.

From all these limitations, imposed by the nature of materials, the designer is now free. Plastics have given him a new and almost unbelievable liberty for experiment and expression.

How will this limitless control of material be used ?

* *The Works of Man*, by Lisle March Phillipps. (Duckworth & Company. Second edition 1914). Chap. XI, pages 293-294.

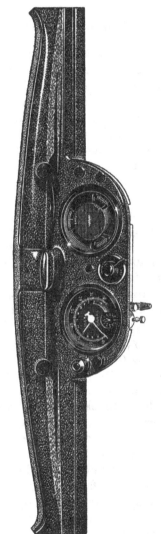

The facia board of the 10-h.p. and 2-litre Sunbeam-Talbot car, made from wood filled urea formaldehyde. This is an example of design aptly catering for a specific function "untroubled by any concessions to preconceived ideas of form", and making the appropriate and effective use of material.

(*The drawing, made from a photograph, is reproduced by courtesy of Sunbeam-Talbot Ltd. and De La Rue Plastics Limited*).

CHAPTER IV

Plastics and the Public

BEFORE venturing to attempt any answers to the question: "How will the limitless control of material be used?" the receptive powers of the public should be considered. The consumer who pays the piper is entitled to call, or spoil, the tune. Often he has it spoiled for him by the manufacturer who has either neglected to employ a designer or has employed the wrong sort. Far more often it is spoiled by that anonymous but exceedingly powerful dictator of taste, the retail buyer, who holds his job because he succeeds in persuading his employers that "he knows what the public wants!"—a bold claim, for he merely knows what the public *may* buy if the price suits them.

But the departmental buyer of the large retailing concern is not the only claimant to intuitive knowledge of what the public wants: there are writers, critics, reformers and other publicists who have theories about this subject, though they seldom realise that what pleases the sensitive highbrow fails utterly to please the factory operative who spends his day in a Yorkshire mill. Very often the products of industry are hopelessly incompatible with the educated taste of the critical minority, while products acceptable to that minority repel the vast mass of the public. That fact is seldom faced by the highbrow critics—the so-called intellectuals. Those exclusive and superior people resemble the seventeenth century puritans, both in their joyless outlook and their moral earnestness. With frigid intolerance they believe that people *ought* to like this, that or t'other, which *they* happen to like. They are impatient of human foibles and they ignore them because they are inconvenient and tend to divert the flow of their militant idealism. It is a national habit to pretend that inconvenient facts are non-existent, but here are some awkward facts about public taste, as opposed to educated or "progressive" taste, that must affect the character of goods made with limitless control of material.

Now it is said that the machine creates its own æsthetic forms; millions of words have been written and spoken in many languages about machine art and what is called "functionalism"; and many sincere and highly educated people believe that good industrial design must always eliminate all evidence of human weakness, for the word weakness is sometimes used to designate the ancient human love of ornament. Machine-made objects should, they contend, be stark and smooth and untroubled by any concessions to pre-conceived ideas of form inherited from the days when things were made by hand. So far as products with a mechanical function are concerned, they are right. If designers are not perfectly frank about the function of such products they soon find that they are disguising, not designing. The history of the radio receiving set may be quoted as an example.

"At first it was a piece of complicated and excessively untidy machinery. The crystal sets of the early 'twenties with their tangle of cables and ear-phones and apparatus, exposed their entrails to the world with the unreticent brutality of a typewriter. Then came refinements of mechanical invention; valves and loud-speakers and all manner of subtleties, and from an engineering point of view, compactness became possible.

"The gramophone and the radio set were presently combined and the radio-gramophone as a composite piece of apparatus displayed such a violent reaction from the crude mechanical exposures of earlier days that it was sometimes difficult to tell, when all the doors were closed, exactly what this new piece of furniture was.

"For instance it might borrow its inspiration from early Stuart or Carolean furniture. It might hark back even further, and an imitation Tudor food hutch might house the latest mechanical triumph of the twentieth-century.

"Again, it might seek a borrowed elegance from the William and Mary period, elevating upon legs a cabinet that contained all the essential mechanism.

"But at last there was a rationalizing process, and the work of engineers was no longer accommodated in cabinets that were earnestly pretending to be something antique.

"The radio-gramophone was acknowledged as a piece of furniture with a function and form of its own. The wireless receiving set, as its apparatus became more compact and made fewer claims on space, began to acknowledge its obligations to truth about its function, and in that acknowledgment it could display simplicity and comeliness, and could provide the cabinet-maker with the opportunity of using decorative veneers."*

The next stage was the radio cabinet moulded from plastics, and

* *Industrial Art Explained*, by John Gloag (George Allen & Unwin, Ltd., 1934). Chap. IV, "Examples of Development in Design," pages 119-112.

Industrial products with a mechanical function demand recognition of that function. "If designers are not perfectly frank about the function of such products they find that they are disguising, not designing." After years of disguise, the radio set received attention from some of our leading industrial designers in the mid-nineteen thirties. The use of moulded plastics furnished an opportunity for new forms; and here is an example of the completely new approach that was then made to the problem. This Ekco radio set was designed by Misha Black.

(The drawing, made from a photograph, is reproduced by courtesy of E. K. Cole Radio Ltd.).

The influence of plastics on the form of a radio set, when a skilled designer
uses the materials to the best advantage, is again agreeably demonstrated by
this Ekco model, designed by Wells Coates, F.R.I.B.A., and first produced in
1935-36. Such forms, when they are associated with specific mechanical functions, are acceptable to the public, in any appropriate material. If this design
is compared with those illustrated on pages 36, 39, 40, 42 and 44, it becomes
apparent that they have something in common, although they may differ
widely in form or function. Their functional fitness "is uncomplicated by any
inherited notions of form or finish: their designers have started clean."

(The drawing is made from a photograph and is reproduced by courtesy of E. K. Cole
Radio Ltd.).

when leading industrial designers like Misha Black, Serge Chermayeff and Wells Coates solved this problem of design with the new materials, they achieved forms as distinctive as those possessed by such instruments as the violin. They were as frank about the function as any maker of violins: as first-class designers, they studied the limitations and possibilities of the materials, and with imaginations informed and stimulated by such study, they produced graceful solutions.

Now, such simple and well-proportioned shapes are generally acceptable when they are associated with some mechanical or semi-mechanical purpose. For instance, the H.M.V. Electric Iron, designed by Christian Barman; the "Aga" heat storage cooker, designed by Dr. Gustaf Dalén; the "Otto" stove, designed by Raymond Loewy and Charles Scott, all exemplify the essential qualities of machine art: the functional fitness of these products is uncomplicated by any inherited notions of form or finish: their designers have started clean. The manifest agreeableness of such pieces of specialised apparatus cannot be accepted as a proof that the public would welcome a universal application of the standards of machine art. There was little indication before the second world war that home-makers, either in the lower income groups or representing the better paid type of factory operative, desired to surround themselves with clean, simple, untroubled shapes. The reason is fairly obvious: a man who works all day long in a factory among angular and busy machines, who spends nearly half his life in the shadow of their implacable efficiency, does not want to be reminded of his work when he reaches home. He wants possessions that have some individuality, things with a flavour of their own. His purchase of pseudo-traditional furniture is a form of escapism. His education has not taught him to observe and compare; his critical powers are uncultivated, nor are they called into action, unless he is confronted with things that have an explicit mechanical function, such as a motor car, a cycle or a radio set.

That was the state of public taste up to 1939. I am not concerned with the rights and wrongs of this: I am concerned with the facts as they were before the second world war began. Such facts should not be brushed aside as unimportant or irrelevant. They should not be briskly disposed of by saying that people ought to want or like something that stirs the emotions of the highbrow reformer. An enormous gulf separates the confident reformers from those whom they seek to reform—the easy-going, good-humoured lovers of sport, cosiness and privacy, the people of England. Although the

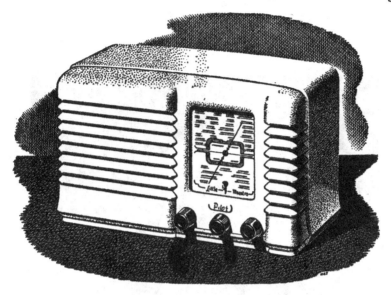

A radio set in cream urea formaldehyde. This is another compact and agreeable solution to the problem of accommodating a radio receiving set. Mechanical function has dominated the form: the materials used have allowed that form to be simple without being stark. Designed for Pilot Radio Limited by M. M. Lynn. (*The drawing is made from a photograph and is reproduced by courtesy of Pilot Radio Ltd. and De La Rue Plastics Limited*).

highbrows may regard themselves as the intellectual aristocracy of to-day, they lack the warm human appeal of the pleasure-loving and artistic aristocracy of the golden age of English design, which lasted from 1660 until round about 1830. As reformers, they proclaim and undoubtedly possess a profound interest in the welfare of the people; but these alleged leaders of taste and thought are without influence and without imitators among the people. And it is the people, the thousands of workers who will be making new homes and re-making old ones after the second world war, who will decide, all unconsciously, the commercial success or failure of goods designed and produced with limitless control of material.

I believe that we have a fifty-fifty chance of a renaissance of public taste; but because we have passed through a period of compulsory utility and austerity, there is also a danger that the public may react so violently against those necessary war-time restrictions, that we shall have an orgy of ornament. I have said that love of ornament is an old human characteristic. It can become

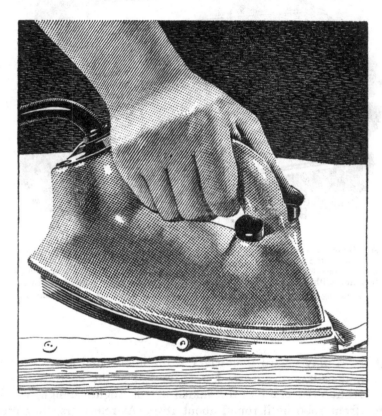

This illustration, and the two that follow, are included to show how articles made in materials other than plastics bear the impress of a common way of thought about industrial design. The radio sets on the previous pages and the cooking and heating apparatus on pages 42 and 44, are, like the electric iron shown above, solutions to a mechanical problem. These solutions "exemplify the essential qualities of machine art" and, again, their functional fitness "is uncomplicated by any inherited notions of form or finish: their designers have started clean." The H.M.V. electric iron is made of hard-glazed fireclay and metal: it is designed by Christian Barman, F.R.I.B.A. (*The drawing is made from a photograph and is reproduced by courtesy of H.M.V. Household Appliances*).

a virtue or a vice. As part of their war effort, most people have learned to want what they could get, though they were not always as conveniently pliant as tidy-minded bureaucrats could have wished. When they can get what they want, they may well be in the mood for cramming into their houses more decorative objects and even more conflicting patterns than Victorian householders. And here lies the danger: these new versatile plastic materials, with their immeasurable capacity for variation in form and colour, may be used to create a new rococo period. Reaction from austerity might have a strange effect on public taste: it could be vulgarised or improved. The facilities furnished by the plastics family could encourage either tendency. But ultimately the public calls the tune; and they are just as likely to call a good tune as a bad one; and industry can provide plenty of good tunes. I believe that they would be popular, and for the following reasons.

We must remember that we shall have a new public: for the first time in our history, thousands of women have been uprooted from their homes, have served in the auxiliary forces on a national scale, and have come into contact with modern machinery used in the war effort, machinery designed to perform tasks with the utmost economy of effort and material. Also, thousands of men have had to manage superb machines in the Navy, Army and Royal Air Force. We all know how incredibly tidy and neat and deft men become when they have served in the Navy, whether as ratings or officers. Such characteristics are also acquired by men who serve in the operational side of the R.A.F., and in the mechanised units of the Army. Contact with well-designed, efficient machinery has a definite effect upon character: this was as true in the days of sailing ships as it is in the days of steamships and motor ships. The sailor acquired the reputation of a handy-man; and I suggest that a big section of the post-war public will consist of critical handy-men and women. The women, after their service experience, will desire privacy, comfort and cheerful surroundings in their homes; and I doubt whether they will tolerate inept, old-fashioned equipment. Women who have helped to operate such an admirable piece of machinery as a range-finder in an anti-aircraft battery, who have helped to make certain types of sea-going craft ready for service, who have helped to build Spitfires and Hurricanes, are likely to welcome light and easily cleaned equipment in their houses, and especially in their kitchens; they are going to appreciate the smooth translucent and gaily coloured plastics that will be available for the making of kitchen equipment and furniture, and they are going to

The "Aga" heat storage cooker, designed by Dr. Gustaf Dalén. The scale of the problem may change, but the approach to design is essentially the same here as with the radio sets and the electric iron shown earlier.

(This drawing is made from a photograph and is reproduced by courtesy of Aga Heat Limited).

enjoy shopping in shops and stores where plastics not only play a prominent part in the devices used for the display of goods, the showcases, stands, lights and labels and so forth, but in the goods themselves.

I believe that both men and women are going to be more critical about the things they use, and those sections of industry that are concerned with the provision of materials and equipment for the post-war home may find that the ultimate consumer has become as critical of the shape and performance of everyday things as he was before the war of the performance and efficiency of his motor-bicycle or car. An improvement on 1939 standards of public taste may thus come about; though it could only be a limited and tentative improvement. What I said earlier about the factory operative desiring to have nothing in his home life which may remind him of his working day would still be true; for although association with well-designed machinery on war service may sharpen the critical judgment of receptive people, there can be no rapid change of public taste. The people at home would remain unaffected. As a result of the war, we may have a nucleus of young people with a new outlook about the design of everyday things; people who as consumers may be responsive to new experiments in design. They may even represent a substantial addition to what an American industrialist once described as "the useless five per cent." When asked to explain what he meant by that extraordinary statement, he said: "The five per cent of the population which knows what it wants, and is therefore no use to industry." The "useless five per cent" has hitherto been represented almost wholly by consumers with comparatively large purchasing power; but if we have a percentage of general consumers in the lower income groups joining the elect brotherhood of critical people who know what they want, we may begin to see a gradual improvement in the taste of the public, and a corresponding increase in the market for well-designed goods. That new, critical awareness should be anticipated, and it should be honoured. To satisfy it, industry should call on the services of our industrial designers, for consultation with industrial designers is one of the operations of industrial production, as progressive British manufacturers have long recognised.

We shall have many ingredients for a new renaissance of taste in this country after the war, and it may bring to the commercial machine age in which we live the gracious values of humanism. We shall have a new, and perhaps more receptive section of the public with young minds from which old prejudices have been

The "Otto" stove, designed by Raymond Loewy and Charles Scott. Like the "Aga" cooker and the H.M.V. electric iron, this stove, which is made largely of cast iron, shows how all materials can be used without complexity when the industrial designer sets out to solve a functional problem and creates the most suitable form to achieve that solution. Compare this design with those shown on the previous pages in this chapter.

(This drawing is made from a photograph and is reproduced by courtesy of Allied Ironfounders Limited).

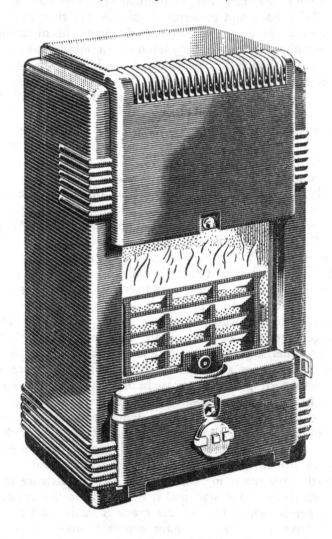

dispelled by experience. We shall have a galaxy of new materials; not only plastics but many others, perfected during the war by the genius of our research chemists and metallurgical specialists. Already we have a new generation of manufacturers with the courage to make experiments in design, and who, in common with most British manufacturers, refuse to put goods on the market until they are satisfied, by exhaustive tests, of their integrity. Also, we have some of the finest industrial designers in the world.

But the productive collaboration of progressive manufacturers with industrial designers is not enough. So far as the home market is concerned, manufacturers are often dictated to by retail buyers. It is essential for the good health of our export trade to have the sustaining influence of a flourishing home trade. What sort of a man or woman will the retail buyer be after the war? Will he or she recognise the fact that there may be quite a new sort of public in the making? Will he or she insist that plastics must be dolled-up to look like wood or metal? If retail buyers are nervously obstructive, they will not only deprive the new young-minded British public of well-designed goods in plastics, but indirectly they will harm the capacity of British manufacturers to compete in world markets.

I sympathise with the retail buyer: he is often haunted by the thought of the accountant in the background muttering threats about departmental sales figures being maintained. But unless we revive, both in our retailing at home and our marketing abroad, the spirit that for three and a half centuries has sent British goods and trading adventurers round the world, we shall lose the opportunities that these new and remarkable materials have brought to British industry. The stimulating effect of plastics upon industrial design could, if properly directed, help to restore our national prosperity. An immense volume of chemical research has been organised by the British plastics industry, and if we are to use these new synthetic materials wisely, it must be recognised that development research work in design has a corresponding importance to the initial work of the laboratory. By using the trained imagination of the industrial designer, we shall be following up to the best advantage the trained imagination of the research chemist. The public may accept any amount of rubbish, but they seldom reject the results of genuine imagination.

CHAPTER V

Industrial Design and Commercial Art

VARIOUS aspects of the nature and possible effect of plastics have now been discussed; in particular their potent appeal to the manufacturer and the designer, and the dangers that may consequently arise from excessive stimulation of the imaginative faculties of those partners in modern industry. The receptive powers and unpredictable taste of the general public have also been touched on; but the receptive powers of designers and their capacity to use these new materials appropriately and inventively for solving various problems of design must also be examined. To avoid confusion or prejudice, some definitions and broad classifications of the activities covered by the word *design* are necessary.

Design is a business operation, whether it happens to be industrial design, which is a basic operation in the production of goods like kettles, cameras, or electric radiators and large-scale objects such as railway rolling stock; or commercial design—commercial art as it is generally called—which affects a number of operations in the task of distributing goods or offering services to the public. Industrial design has a considerable, though often unrecognised, influence on the things that people buy and use: commercial art has often played a decisive part in bringing people to the point of purchase, though the purchaser is seldom consciously aware of the existence or function of the commercial artist.

Industrial design becomes a basic operation in the production of goods when trained imagination is introduced to secure the most efficient, agreeable and inventive use of appropriate materials and processes. The industrial designer is the man with the trained imagination; he is a technician, and his work is just as basically important as that of the research chemist or the production specialist. Omit the designer, and you have omitted imagination, and very often lost the chance of gaining a new market or reviving and extending an old one.

There are two divisions of industrial design, and I put them in this order of importance:

1. Design which is concerned with the *form* and *function* of a manufactured object and which determines the selection of materials

and fabricating processes. It is for this division that the industrial designer works primarily, though his interest and activity are often carried into the second division.

2. Industrial decorative art, which is concerned with the creating of *decorative patterns* and the use of colours and textures in relation to such patterns. Many capable artists are attracted by the opportunities this division affords, though it should be understood that creative ability successfully expressed in the practice of industrial decorative art is not necessarily a qualification for the far more exacting needs of the first and principal division. Industrial design demands a different, though no less imaginative, type of mind.

These divisions are suggested to me by the practice of the National Register of Industrial Art Designers, who classify work under two headings: (*A*) Designers of *Shape*. (*B*) Designers of *Decoration*. I will deal with the use of plastics in connection with *commercial art* after the two divisions of *industrial design* have been examined.

The first division of industrial design covers an enormous range of goods as well as large-scale objects, ranging from motor-buses to trans-Atlantic liners. The term *light engineering* has sometimes been used instead of industrial design to describe the production of goods that have some mechanical function; but that term apparently suggests to many people, no doubt erroneously, an excessive pre-occupation with metal. Industrial design embraces the use of all materials, organic and inorganic. Consequently, an industrial designer must have the capacity for studying materials and the training which will assist him to apply such study in a practical manner. Some of our finest industrial designers are architects, for their training enables them quickly to grasp the limitations and possibilities of production processes without disabling their imaginative prowess. Through the medium of industrial design the architect may again assume the universal responsibility for design that he enjoyed in the eighteenth century. In that age the architect was the man who could, and indeed did, design anything from a coach lantern to a watch chain, from the decoration on the stern gallery of a man-o'-war, to the planning and complete equipment of a mansion for an affluent member of the nobility or gentry. I believe that if the young post-war generation realises that industry offers limitless opportunities to the architectural profession, we shall get some of the best trained imaginations in the country behind the products of British industry.

The industrial designer who is concerned with a three-dimensional problem of design, whether it is some mass-produced article of domestic equipment or the "tidying-up" of a liner's superstructure, may be relied on to study the diversity and to become acquainted with the properties of the various members of the plastics family. Obviously, he could never know as much as the chemist about plastics; but he may often perceive qualities that are hidden even from the most imaginative of research workers in a laboratory. The designer looks at a new thing with new vision.

For example, the research chemist produces new synthetic materials; but he often circulates specimens of his experimental productions that do less than justice to their range of properties. He is content if he demonstrates the nature of the structure, the strength, the weight and so forth. In the course of conducting design research work, I have seldom found much appreciation of colour emerging from the laboratory; and many a new experiment in plastics makes its first public appearance in some repellent shade of brown or grey, resembling a slab of stale bread or a lump of putty. But the designer looks beyond the uninspiring samples of material he is sometimes shown when he begins his study of plastics: he wants to know all about colour and texture, and the possible variations of hue and surface treatment, even though his initial concern is not decoration.

The industrial designer, particularly when he happens to be an architect, is used to dealing with technicians; as a master technician himself, he wins their respect. So when he is one of a team, composed of production engineers, research chemists and sales executives, engaged in design research work, he enjoys a receptive atmosphere directly his standing as a technician is acknowledged. He can often make suggestions to specialists which would be resented, or at least discounted, if they came from a sales executive; for a ridiculous and needless hostility sometimes exists between the sales executive and the specialist with scientific training. The industrial designer is often the solvent for such suspicious obstructiveness, for he has an equal interest in the production and sales aspects of a problem. When industrial design is recognised as a business operation, and moreover a basic operation in the production of goods in a factory, it becomes properly related through the sales organisation of a firm to the market that represents the manufacturer's ultimate objective. I have described, in a book devoted to the organisation of development research work in design, the technique of the design research committee or panel.* This particular solution to the problem of

* *The Missing Technician in Industrial Production*, by John Gloag. (George Allen & Unwin, Ltd., 1944).

practical collaboration between industrial designers, production specialists and sales executives has been adopted in many industries; it is by no means the only solution; but it is particularly effective when exploratory work with new materials is essential. The following example may be cited from the book just mentioned. "Several years before the war, in carrying out advisory work on industrial design for a firm that was manufacturing bathroom equipment and small articles of furniture, I helped to organise a Design Research Committee. The work of that design committee was the ground plan, so to speak, of a good many similar committees which I have since been instrumental in forming. This committee was concerned with three materials: timber, plastics, and glass. Plastics in those days were rather an unknown quantity . . . and the job of the industrial designers was to evolve the most convenient form for a variety of articles such as bath trays, coat hooks, and tooth-brush racks. The first drawings were made by the design members and discussed at the committee table with the plastics engineers and the sales executives. The question of the practical production of the shape suggested was thrashed out, calculations were made regarding economic runs and finishes, and the designers in consultation with the plastics technicians could, after being briefed in this practical way, modify a design and recast it to secure the best results from process and materials. At that preliminary stage, all questions arising from the initial cost of moulds, the type of plastic to be used, and the possible market, could be discussed. The drawings were then revised, and from them plaster or wood models were made, so that, in this third-dimensional stage, final adjustments and refinements could be settled. In this way, the best design for expression in terms of new and rather strange materials was secured. Before this committee was formed, it had been usual to select some familiar article made in wood or metal, and to hand it over to the plastics experts, to copy in their material."*

The work of that particular committee completely changed the approach of the firm to the use of plastics: they were never again regarded as substitutes. I have been associated with other design research committees which have explored the use of plastics in a variety of fields, and have thus had many opportunities of observing that the approach of industrial designers to a problem, when they are properly directed by specific terms of reference, is distinguished by clear and original thinking. They are never intimidated by a

* *The Missing Technician in Industrial Production,* Chapter III, pages 32-33.

prototype, evolved with older materials: they are never afflicted by the mental indolence which mistakes mere novelty for inventiveness.

I have seen some experimental types of furniture made from various plastics, which disclosed the absence of a designer or the laziness of a designer, if indeed one was employed. In some examples, chairs of traditional form had been made in a transparent plastic, so that shimmering ghosts of Queen Anne or Chippendale models had unhappily materialised; in others, chair frames were constructed from transparent rods, bound together in imitation of the technique used for constructing cane furniture. An air of slightly uneasy luxury was conferred upon these articles, because transparent materials possess a luxurious quality, but they were destitute of original inspiration in design; true, they were unusual, but only in the way that Cinderella's glass slipper was unusual. Such things could only be produced by people who were still in the "substitute" stage of thinking about plastics.

The reaction of the industrial designer to the manifold gifts of the plastics family, is first to discover how stresses and strains, tensions and weights are affected by the materials; what limitations on shape and size are imposed by the various fabricating methods; whether surfaces are resistant to wear and tear and what upkeep, if any, they require. Equipped with such knowledge, he tackles with an open mind the problem of designing, say a chair. He only has to accommodate the contours of the human frame, and although the posture adopted by most Europeans when they are seated has changed slightly during the last century, we are still vertebrates, even though we no longer care to sit bolt upright, as our great-grandfathers did, and lounge a few inches nearer the floor than they would have thought consistent either with dignity or decency. Since the early sixteenth century, chairs in England have been supported by legs; before that they were boxes with a high back and solid arms rising above the top of the box. Woodworkers learned how to economise in their material, and to trust its strength in new ways; so four legs, linked and braced by an underframe, eventually supported the chair seat. Then the underframe was eliminated, and the legs were tapered. For a couple of centuries a progressive refinement of design gave increasing elegance to the chair; then the slimming process stopped, and throughout the Victorian period the chair grew bloated and sank down to the floor; its legs bulged and thickened, its swollen feet were shod with castors. Now, the industrial designer has the problem of making a comfortable seat with "limitless control of material." The result may be a two-piece chair: a back and a seat in one piece, and a curved underframe in one piece to support it. That is, perhaps, an excessively simple solution; but it is not an impossible one, and I have mentioned it to illustrate the liberation of design from ancient limitations and pre-conceived notions about the form of anything that occurs when an industrial designer is working out a problem with plastics.

The second division of industrial design is complicated by the existence in this country and elsewhere of a vast number of so-called specialists in decoration. These enthusiastic people, many of them without taste or training, grasp at every opportunity for ornament, and once they realise the prolific capacity of plastics for enlivening surfaces and making blobs and knobs and wriggling shapes, they will insist on giving us all the colours of the rainbow plus all the confusion of "the morning after." But there are many able designers whose gifts can be productively engaged in the study of plastics in connection with industrial decorative art. In the devising of patterns for the surface decoration of some types of plastics their judgment and skill would be invaluable. For instance, there is a type of plastic which is built up from layers of paper, textiles or even thin veneers of wood. The layers are impregnated and bound together with a phenolic resin.* These laminated plastics may be plain in colour, with smooth, hard, glass-like polished surfaces; or a satin finish or texture may be imparted to the surface from patterns stamped on the metal sheets of the press wherein the sheets are formed. Patterned paper or textiles may be used for the top layer of the lamination, or the plain surface may be ornamented by cutting out inlays through the sheets before compression.

The trained designer will know how to handle such opportunities; though the judgment of the untrained designer may, like some mediæval English monarchs, die of a surfeit. For interior decoration, for lighting fittings and ornamental articles, bowls, vases and lamp-shades, plastics have already become established in use; and in Chapter II, Mr. C. F. Merriam's reference to their extensive employment on the Cunard-White-Star liner, *Queen Mary*, was quoted. Apart from interior decoration, the future of plastics in the fashion world carries us beyond the consideration of industrial design. To the designers of such things as costume jewellery, fancy buttons, buckles, handbags, compacts, shoes, belts—all the appurtenances of feminine toilet—plastics bring a new treasury of ideas. Again, the work of textile designers must in time be profoundly

* See Section II, pages 90-92.

affected by the increasing use of specialised types of plastic fabrics. Like the progressive potters, textile manufacturers have worked out many fruitful partnerships with designers, and the study of plastics must inevitably initiate fresh design research work in their industry.

The mastery of decorative industrial art attained by some of our artists and designers in ceramics and glass, suggests the richness and variety of talent that could bestow comparable distinction on things produced in the new "fifth class" of materials. This mastery is exemplified by the work of artists like the late Eric Ravilious for Wedgwood pottery; by Keith Murray's work for the same firm, and his designs in domestic glass for Stevens and Williams Limited; by the decorative treatments carried out for Pilkington Brothers Limited on various forms of glass, by Kenneth Cheesman, Sigmund Politzer and Hector Whistler, and by the patterns for glass designed by Paul Nash and R. A. Duncan for Chance Brothers and Company Limited. Only a few names have been mentioned of men who excel in this field: there are many others.

Having briefly examined the two divisions of industrial design in relation to plastics, commercial art may now be considered. Commercial art is not concerned with the form, function or production of goods, but only with their presentation to consumers. It has, perhaps, greater vivacity and certainly less permanence than industrial design. The term covers all those branches of artistic activity which assist the distribution of goods, such as press advertising, posters, booklets and leaflets, the labelling, packaging and display of merchandise, the design of display material in shops and stores, and the design of exhibition stands. It commands the services of commercial artists, typographers, packaging and display specialists; and to some of these people plastics will furnish opportunities for original experiment. It is obvious that in the packaging and display of goods, there will be an increasing use of plastics, particularly of transparent and translucent varieties.

Before 1939 there had been a great increase in the use of plastics for packs, for stoppers on bottles and tins, and for wrapping. But added gaiety and dramatically luxurious effects for the packing of such things as toilet preparations only represent one aspect of the service that can be rendered by plastics. Here is an example of the way the form and character of a container could be changed by an imaginative designer. The problem is presented by very small tablets of a drug, usually contained in a bottle which can be slipped into the waistcoat pocket; but the minuteness of the bottle causes a difficulty, for it only allows two restricted surfaces for accommodating a label, and on that label it is essential to set forth explicit directions for taking the tablets, their ingredients, and the name of the makers. This information has to appear in four languages, and even the use of a label that encircles the bottle cannot solve the typographical problem satisfactorily; so perforce it is solved unsatisfactorily, and the smallest type that can be set by a compositor is reduced still further by photography, until the essential material is crammed, illegibly, into the space available. The use of a transparent plastic would allow a packaging designer to produce a disc pack, smaller in diameter than a watch case and not much thicker than, say, three half-crown pieces. This disc pack would have a simple, screw-thread closure; the two sides of the disc would offer a much larger area than the bottle for printing the directions and so forth, and possibly the material could be printed direct upon the plastic surface.

A convenient form of pack for tablets has been in use for some years, consisting of sheets of transparent material, enclosing the tablets in parallel rows, so one or more may be torn off easily. Creative thinking by commercial artists and packaging designers may change the manner in which all kinds of goods are delivered to consumers, and shopping in the future may not only be more exciting, but far more convenient.

The growing influence of plastics in the display of goods was demonstrated by the New York World Fair in 1939. I observed, both in New York and in the San Francisco World Fair, which I visited in that year, a fresh approach to display problems which suggested that some remarkable ideas about the alliance of light with transparent and translucent materials were encouraging designers of display to make experiments. The enclosing of objects in irregular masses of transparent plastic, so that they interrupted and redistributed a beam of tinted light; the bubble delicacy of barely discernible display units, allowed goods to be arranged on almost invisible shelves and supports; an infinity of variations for edge lighting, with artificial light transmitted from edge to edge of transparent sheets of polystyrene*—all these and many other partnerships between plastics, glass, metal and light were then being explored by American designers. The significance of such work had not of course escaped British designers; but the war interrupted developments.

As I said in the opening chapter, it is the purpose of this book briefly to examine plastics in relation to industrial design: to

* See Section II, page 80.

attempt even a casual review of the ramifications of commercial art is beyond its scope. The subject has only been touched on in this chapter to avoid any possibility of the functions of industrial designer and commercial artist becoming confused. I do not imply that they are rigidly segregated: technical skill illumined by imagination will always transcend the tidy docketing of official registers and academic institutions. By virtue of training and creative ability the same man may be an industrial designer and a commercial artist; obviously no label, used merely for convenience, can affect his gifts or knowledge when he is solving a problem; but if the problem he solves is concerned with the shape, function and finish of goods, then he is practising industrial design; if his talents are exerted for devising methods of packing or exhibiting those goods, then he is practising commercial art.

If our designers are allowed to do their best work in collaboration with manufacturers who are making goods wholly or partly from plastics, the public at home will certainly be confronted with the results of genuine imagination, and we shall woo markets abroad with the help of a vigorous salesman whose salary never appears on the pay roll of a sales department.

CHAPTER VI

Economics of Design

UNLESS all the partners in the operation of industrial design acquire a realistic outlook about the facts of commercial life, the character of goods made in Britain is likely to be impoverished and ill-adjusted to the standards, the taste and the expectations of the new period of social and economic life into which Western civilisation is slowly and painfully passing. It has been said, with the confidence which inspires eminent politicians when they are uttering sonorously vague words, that this will be "the century of the common man." It may easily become the century of the commonplace; for undue sensitiveness to what the common man is *supposed* to find acceptable is but an inverted form of the arrogance which impels the highbrow to assert that the public *ought* to like what *he* happens to like, while passive acceptance of what is presumed to be popular taste leads to the adoption of "safety first" as a manufacturing policy and ultimately to the

dissolution of enterprise. Manufacturers, industrial designers, production specialists and sales executives are all partners in modern
industry; but they are unlikely to abdicate leadership if they accept
industrial design as a business operation.

Once we enjoyed world leadership in industry: the way our
goods were made, the excellence of the materials we used, the skill
and inventiveness that enriched the character of most British products secured for us a remarkable reputation in the iron, steel and
steam phase of our industrial development. The economic necessity
for good design was not then so pressing; we were the world's
master machine-craftsmen, though the opportunity for becoming
the world's master designers was missed, because nobody realised
that such an opportunity existed. The word *design* in the nineteenth
century generally signified ornament; that unfortunate mistake still
confuses the ideas of some manufacturers; but unless they recognise
design as a business operation, they are unlikely to remain manufacturers in the new phase of our industrial development, which
began after the first world war and since 1939 has been intensified.
Steam, iron and steel dominated the industrial technique of the
last century; electricity, plastics and light alloys now form a new
association of power and materials whose economic employment
requires a corresponding association between design and industry.

In the last chapter it was said that when industrial design is
recognised as a business operation, and moreover, as a basic operation
in the production of goods in a factory, it becomes properly related
through the sales organisation of a firm to the market that represents
the manufacturer's ultimate objective. Many obstacles are placed
in the way of industrial design gaining such recognition, and some
of them are raised by people who are passionately eager to improve
design but cannot think of it except as a form of uplift or as a
symbol of so-called progressive political faith. If the technical and
business operation of industrial design is regarded principally as
a means of reforming obtuse or unreceptive manufacturers, and
securing their conversion to some brand of highbrow belief about
life and art; if indeed, industrial design is thought of as anything
but an essential part of commercially successful practice in the
production of goods, then those who advocate it because they
believe in good design for its own sake are doing a grave disservice
to their chosen cause. Such people, whose sincerity is indisputable,
constantly attempt an over-simplification of economic and industrial
problems. It is easy to say: "The whole economic system is rotten
and must be scrapped!"—easy and lazy. Reformers and critics

who use such dismissive phrases are seldom prepared to consider methods of improving existing industrial organisation. They clap a particular brand of political faith on a problem—a sort of red-hot mustard plaster—and leave it there, hoping for the best, believing the worst and ignoring the facts. It is usually an article of their faith that British industry is greedy and inefficient, "Blimpish" and bone-headed. If, with hesitant common sense, you suggest that British industry is not only alive and kicking, but is sustained by a spirit of high adventure, you are labelled as a reactionary by the change-everything-at-any-price type of reformer. But whether they approve or disapprove, the reformers, like everybody else, live in the commercial machine age, or "the scientific commercial age," if we adopt a Wellsian description quoted earlier. In this age industries must pay their way, and though that fact is disregarded or despised by many of the superior people who desire to improve standards of design, an industrialist cannot afford to forget it, so he naturally asks whether industrial design is an economic necessity: in short, does it pay?

Other questions arise out of this, for the manufacturer who has produced goods without ever consulting a designer, may ask: is industrial design even necessary? Well, the external application of water is not necessary in order to support life: it merely makes people more agreeable. Industrial design is not necessary in the production of goods; but it makes goods more acceptable. Our life as an industrial nation may depend upon our ability to recognise this fact, for there is a new spirit abroad in the world—a spirit of expectancy. People everywhere are expecting something vaguely better from the post-war world than they have ever had before: they are encouraged by the discreetly vague hints of statesmen, by the ardent anticipations of the press, and by the feeling that they deserve to live in an easier, more pleasant world. If we go on producing goods and omit the essential ingredient of imagination, we shall not be able to give people anything that reflects the power and capacity and thought of the mid-twentieth century, nor shall we be accepting and fully using the new power and materials phase of world industry.

"All very well," the cautious and conservative manufacturer may say, "but I've enough good bread-and-butter lines to keep me going." But in the atmosphere of the new world the bread may soon become stale and the butter rancid; for many people who are comforted by the existence of "bread-and-butter" lines forget that they are the result of yesterday's enterprise and experiment, and that unless a business is continually refreshed by enterprise and experiment it has no to-morrow.

"Ah, but it's all a question of price," the manufacturer may continue; "if the price is right, what more do you need?"

Well, to begin with, you need a lot of confidence in the lassitude of your competitors; for some other firm may produce an article which competes so actively in appearance and convenience with your goods, that it walks away with the market, even if it is sold at a higher price, though it is quite likely that as a piece of good industrial design it may be produced more economically, so that you are undersold as well as outmoded.

In conducting research work in industrial design, I am occasionally asked: Are plastics going to be cheaper than—this, that or the other material? I don't know the answer to such questions: very few people would venture even on a rough guess. But the question shows a fundamental lack of appreciation of the potency of good design as a sales factor in the production of commodities. Good design forms a broad bridge between raw materials and consumer needs, capable of carrying an enormous press of traffic; and the excellence of the bridge increases the volume of the traffic. In June, 1941, the technical magazine, *Plastics*, published an unsigned article entitled: *Prime Cost Bows to Design*. Under the heading of *Cost and Value Relationships* the following paragraph appeared:

"Generally speaking, we insist that a new design, a new idea and greater stability of purpose are fundamentally more important than raw material cost. The mentality that utilizes the formula, 'What's the price?' first, conveys, to us at least, the impression of a mentality devoid of ideas and too ready to work down to a price, one which was the cause of price-cutting wars of the past, and eventually resulted in the use of cheap and poor raw materials and skimped workmanship. Without purporting to be moralists our whole structure of production must be built on the idea that we should make a thing better than the other fellow, rather than that we must make it cheaper. If the better thing is as cheap or cheaper, so much to the good, but the two properties are not indissolubly hinged one to the other."*

I don't doubt our national ability to design things "better than the other fellow," if the native genius of our industrial designers is productively engaged. The economic case for good industrial design has been made forcibly and with a wealth of practical examples by the American industrial designer, Raymond Loewy,

* *Plastics*, Vol. V, No. 49, pages 114-115.

in a paper entitled *Selling Through Design*, which I had the privilege
of reading on his behalf before the Royal Society of Arts in December
1941.* He pointed out that "advertisers talk of 'eye-appeal,'
manufacturers of sales and costs, artists of æsthetic values. The
designer assumes the attitudes of all three when he works. His
record in the past twenty years is a triumph of diplomacy, prac-
ticality, invention."† Mr. Loewy is speaking of the triumphant
record of industrial design in the United States: we cannot look
back upon any comparable achievement in this country, and unless
we use our industrial designers we shall have little to look forward
to in the future.

"Yes, but," the manufacturer proceeds, making one more bid to
preserve his peace of mind, "this is like the Fat Boy in *Pickwick*
saying, 'I wants to make your flesh creep!' Can you say *when* good
industrial design pays in terms of sales?"

Good industrial design can often command big or at least
satisfactory sales, but not always. Sometimes good industrial design
fails and vulgar and unpleasant design is highly successful. But I
would never say that because a thing is intended for a cheap market,
that it must be nasty. To couple the words *cheap* and *nasty* is as
fundamentally silly as coupling the words *dear* and *nice*. Everybody
knows that some so-called luxury products sold at big prices are
revolting in appearance and futile in use.

Good industrial design generally pays in terms of sales when a
functional problem has been adroitly solved, and when the article
that is being manufactured is partly a mechanical one. For example,
if any manufacturer of lawn-mowers took the trouble to re-design
that clumsy and needlessly complicated piece of mechanism and
called in an industrial designer to collaborate with his own engineers,
we might be able to buy a lawn-mower that was easy to clean, easy to
adjust, and agreeable to look at. The lawn-mower has replaced the
scythe. The safety razor has replaced the old cut-throat razor. The
safety razor has been designed: the lawn-mower has merely occurred.

Many appliances have demonstrated the commercial success that
follows excellence in design: two outstanding examples are the
Murphy radio set, and the H.M.V. electric iron.‡ But although
bad design also pays, I doubt whether a badly designed object
that had a mechanical function could continue to sell in competition
with something that was well-designed.

---

* It is published in the *Journal of the Royal Society of Arts*, No. 2604, Vol. XC.
  January 9th, 1942.
† *Ibid*, page 94.
‡ See page 40.

Many manufacturers may be tempted to venture a little way into the plastics industry in the future and attempt to do their own fabricating, in the hope that they may not only solve some of their own production problems, but reduce overhead charges by taking outside orders at nominal rates to keep their moulding plant busy. They will buy their experience dearly. A much better and far less costly investment, would be to retain the services of two or three first-class industrial designers, and conduct some development research work in design, with particular reference to the use of plastics, so that the possibilities of these new materials become imaginatively related to the markets the manufacturer hopes to satisfy. I suggested in Chapter IV that in the near future we may experience readier consumer response to good industrial design: I believe that, all over the world, we could become accepted as leaders of industrial design as we were once accepted as leaders of industrial technique. This belief is not based on the sands of national pride or easy optimism: it is built on knowledge of the capacity and skill of British industrial designers, and respect for the courage and enterprise of British manufacturers. We may be entering an age of economic and social turmoil; but it is also an age of opportunity, when new materials, new mechanical processes, new thoughts and new minds may liberate on a new and splendid scale our gifts for design and industrial production and give world-wide scope to our national talent for trade.

CHAPTER VII

The Anonymous Asset

NEVER before have we had such a chance of becoming completely independent of prototypes and pre-conceived notions about industrial design. The second industrial revolution, which was started by the work of men like Michael Faraday and Alexander Parkes, is now developing apace. The rapidity and scale of that development is partly disclosed by the next section of this book; I use the word partly, because plastics only represent one aspect of the new revolution: other materials are growing and changing and acquiring new attributes. How then are we, as a nation, to make the best and the most of the opportunities that are everywhere arising, particularly in this new and dazzling field of industrial activity?

Let us first sum up the various aspects and prospects of plastics in relation to industrial design which have been discussed in this section:—

1. Plastics are not substitutes but are materials in their own right.

2. Plastics do not provide the appropriate answer to every problem of industrial design and production.

3. Their properties and variety bring new freedoms to designers.

4. The limitless control which the designer can exercise over them is a severe test of his imaginative powers.

5. The need for design research work is imperative.

6. Such research work should be conducted by technical specialists in production, chemists, sales executives and industrial designers, working as a team.

7. Such research work must be directed by specific terms of reference.

8. Industrial design, to be effective, must be regarded as a normal business operation.

9. Industrial designers must be regarded as technicians.

10. A new public, attuned to change and receptive to new ideas, may be growing up and establishing itself in the post-war world.

11. People everywhere are expecting, and are being encouraged to expect, a more agreeable environment.

All this suggests the need for fresh and original thinking by the partners in British industry. Fresh and original thinking is not easy; but the lack of it has saddled us with many inconveniences in the past, and even to-day we are surrounded by survivals of the mental blindness that often misdirected the first industrial revolution. For example, that great triumph of the nineteenth century, the railway, was shackled from its inception by out-of-date thinking. The gauge adopted for the track was that of the horse-drawn cart; the ponderous early locomotive showed how much needless weight it had inherited from its parent, the steam pumping engine; the carriages it hauled were stage coaches joined up to form compartments, mounted on under-frames and running on flanged wheels. The motor-car began as a "horseless carriage": only recently has it outgrown the disadvantages imposed by such ideas.

The second industrial revolution could release us from the heritage of inconvenience and inefficiency that has accumulated from lost opportunities in the last hundred years: it could improve or impair the potency of our power to produce goods that would sell in the markets of the world. Without research work in design, we cannot possibly make the best use of such materials as plastics when we are planning the production of commodities. The wisdom and experience and technical knowledge of the industrialist, the chemist, the engineer and the other production specialists are not enough: without the lively, co-ordinating imagination of the industrial designer, without his capacity for innovation, without his constant challenge of accepted and conventional ways of doing this or that, we shall lose the prosperity and national stimulation that the new industrial revolution could provide. We did not lose the prosperity that accompanied the first industrial revolution, because from its beginning and for a century afterwards we were the only highly industrialised nation. We no longer enjoy that monopoly; but we retain the skill that gave us our international reputation as manufacturers. That skill could be enormously amplified by the new industrial techniques and materials; and to direct it the industrial designer can give something that never appears in written words on a balance sheet, but which represents the greatest asset of British industry: imagination.

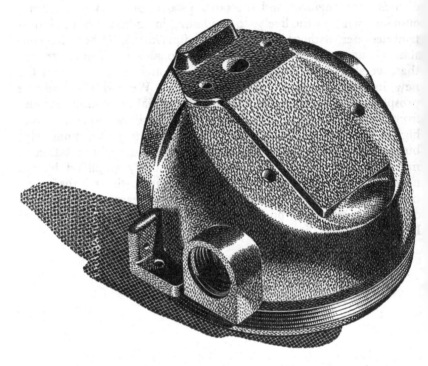

Part of a miner's head-lamp moulded from powder. The making of such a moulding requires the highest degree of skill from the designers of the mould and from the moulders who operate it. This moulding is yet another demonstration of the powers conferred by plastics upon designers and industrialists: it should be compared with the examples illustrated on pages 28 and 31.

(This drawing, made from a photograph, is reproduced by courtesy of Halex Limited).

SECTION II

The Different Types of Plastics, their Properties and Uses

by

GRACE LOVAT FRASER

―――――――――

"Chemists can now make wool from milk, silk from coal, and from such abundant raw materials as wood and petroleum they can produce a greater variety of useful products than exists in nature."

From a paper entitled "Post-war Building — the Chemist's Contribution," *read by the President of the Royal Society of Arts, Dr. E. Frankland Armstrong, F.R.S., before a joint meeting of the Road and Building Materials Group and the London Section of the Society and the Institution of Structural Engineers, September 23rd,* 1943. ("Chemistry and Industry," *February* 12*th,* 1944, *page* 60.)

Introduction to Section II

IN the first chapter of Section I, Mr. Gloag points out that he is not a technician, and I hasten to make the same confession. My knowledge of plastics is not the technical knowledge which a chemist or a production engineer could command: it is based on the study of plastics which I have been able to make in Britain and the United States, thanks to the generous facilities provided by many manufacturers of plastic materials and fabricators of those materials. In 1941-42, I conducted an investigation of plastic developments in the United States on behalf of Halex Limited, who sent me to America for that purpose. My thanks are due to Mr. C. F. Merriam, the Chairman of Halex Limited, and to the directors of that organisation, for permission to set down much of the information that I acquired during my visit to America, and my debt to my American friends in the plastics industry is considerable. In this Section, I have attempted to make a brief, highly condensed and orderly survey of the different groups of plastics, so that it will be easy for the reader to find his way about and quickly to identify any individual plastic and to grasp, with the minimum amount of reading, its essential characteristics and capacities.

G.L.F.

THE DIFFERENT TYPES OF PLASTICS, THEIR PROPERTIES AND USES

SECTION II *by* GRACE LOVAT FRASER

IN this brief survey of the different types of plastics and their uses, a short description is given of the chief plastics within each group, their special characteristics, mode of fabrication and general uses. There are other plastics besides those described, but many of them are experimental and not yet fully proven materials, or are produced for some limited or highly specialised use.

Experimental and development work in plastics is increasing, and war production has given such work additional impetus. The plastics reviewed here bring the picture of industrial and commercial uses up to date.

Plastics are divided into three main groups: thermo-setting; thermo-plastic; and protein plastics. Each of these groups contains a number of different plastics, each with its own particular properties, advantages and special uses. Again, each plastic within the separate groups *may differ within itself*, so that by a change in the process of manufacture it can be transparent, translucent or opaque, hard and rigid or tough and resilient. While the different members of the groups are dissimilar and are made from a wide variety of organic and chemical materials, they are linked to their own group by one common factor.

(1)   In the thermo-setting group this factor is, that the plastic, after being formed by heat and pressure, remains hard and unchangeable. Any alteration to the form of an object fabricated in a thermo-setting plastic can be accomplished only by machining —sawing, cutting, drilling, and so forth.

(2)   The common factor in the thermo-plastic group is that at varying temperatures, softening takes place in the plastic, without a chemical change. Thus, plastics in this group can be reformed if desired; all scrap material can be reground and used again, and the plastics can also be easily manipulated after heat-softening, by bending or twisting.

(3)   The common factor of protein plastics is, as their name implies, that their chief ingredient is of a protein nature.

THERMO-SETTING GROUP

PHENOL FORMALDEHYDE (Moulding type)

British Trade Names — Bakelite: Elo: Epok: Mouldrite P.F.: Nestorite: Rockite.

American Trade Names — Bakelite: Durez: Durite: Heresite: Indur: Makalot: Noreplast: Resinox: Textolite.

Characteristics — Resistant to heat, water, organic solvents, acids and mild alkalies: good low tension electrical properties: greater co-efficient of expansion than metal: some tendency to discolour with time in the paler shades of colour.

Available Forms — Moulding compounds (granules and powder): resin for lamination: blanks cut or punched from a combination of phenolic resin and rag fibres. The moulding compounds are supplied in a large number of different types, each of which is adapted to some special purpose such as shock resistance, high frequency insulation, transparency, etc. They are made from the basic phenolic resin and "fillers" such as wood flour, cotton flock, mica, clay, etc. The blanks are supplied as sheets or stock shapes which can be used either as complete mouldings or as a reinforcement for a part of a moulding where greater resistance to shock is desirable.

Fabrication Methods — Compression, transfer, and extrusion moulding.

General Applications — The variety of special purpose moulding powders available, and the fact that the rate of curing can be adapted to meet almost any industrial need, give these plastics a wide field of utility. The colour range of the general purpose types is fair; the mouldings produced for industrial use are limited to black and the darker colours. Colour stability, except in the dark shades, is not as good as in some other plastics of the thermosetting group. Their toughness and strength, electrical properties and resistance to corrosion make them of prime importance for industrial applications. As their co-efficient of expansion is greater than that of metals, metal inserts can be successfully used with them, the plastic parts gripping the insert firmly as they shrink around it after moulding.

In the electrical field, they can be used for most ordinary applications, except where they are liable to be exposed to ultra high frequency. Switch-cases, plugs, batteries and thermostats are typical electrical uses. They are also much used for radio, vacuum cleaner and typewriter housings; radio bezels and control knobs are frequently moulded from this plastic. In the automobile industry their use is chiefly for switchboard and instrument panels, control knobs of all sorts, door and window handles, and mouldings and beadings for door and window trim. These uses are also extended to aircraft construction.

Though the principal uses of phenolic plastics are industrial, there are also many general applications such as door furniture, domestic electrical fittings, electric iron housings, washing-machine blades, bottle closures, trays, etc. Among a large number of specialised wartime uses were included: tubes to hold the powder charge in demolition bombs, and moulded spigots for portable field water-bags. Wherever high impact strength is needed, plus their other special characteristics, phenolics are indicated and used for a multitude of war purposes. Phenol formaldehyde resins are also used as bonding agents for lamination. (*See* plastic laminates, page 90, and bonding agents, pages 98 to 103).

A recent development is the use of long-cut fibres impregnated with basic phenolic resin instead of the usual moulding compound of resin and filler. The fibres, which may be jute, rami, sisal or kindred types, are impregnated with the resin, pre-formed and compression-moulded. Plastic objects made thus are extremely light in weight in relation to their size, and are of great strength: large objects such as tractor seats, bucket seats for cars and entire car body panels, can be moulded in one piece. This plastic is economical, for only thirty per cent of resin is used to seventy per cent of long-cut fibre. It can only be produced in black or dark brown, but its surface holds plastic paint more readily than most metals, so that any desired colour can be applied, giving a practically permanent finish.

Successful experimental refrigerator and washing-machine housings have also been made from it.

Another form of phenol moulding compound of an inexpensive type, has been developed by the U.S. Department of Agriculture; a variety of agricultural residue products can be used in its manufacture. This compound is formed with twenty-five per cent phenol formaldehyde, fifteen per cent oleo-resinous plasticiser, and sixty per cent agricultural residue. The strength and water-absorption of this compound closely approach those of the ordinary general purpose phenolic moulding compounds; but the colour range is limited as yet.

The agricultural residue products which can be used in this way include: wheat straw, rye straw, oat straw, barley straw, flax straw, rice straw, corn straw, corn cobs, oat hulls, rice hulls, cotton-seed hulls, tobacco stalks and stems, and peanut shells. Satisfactory moulded objects have been made from this compound; its chief use might be for the moulding of industrial objects where a low-cost plastic is desirable. Similar types of phenolic compound have been produced with a bagasse base, both by the U.S. Department of Agriculture, and independent firms.

See Section III, plates VIII, IX, X and XI.

PHENOL FURFURAL

British Trade Names — None.

American Trade Name — Durite.

Characteristics — Rigidity when hot: superior moulding qualities: permanence of dimensions and freedom from scorching or burning at high moulding temperatures: chemically inert: water-resistant: first-rate electrical properties.

Available Forms — Moulding compounds with various fillers: liquid resins: fusible resins and varnishes: modified resins.

Fabrication Methods — Compression and transfer moulding: laminating.

General Applications — These are entirely industrial and electrical, and of the same type as those for phenol formaldehyde. The chief advantage of the plastic is its stability and freedom from burning at high-moulding temperatures; this permits of hot pulling which greatly increases the speed of output from moulds or dies. Its very high heat-resistance makes it an ideal material for such uses as oven thermostats and distributor heads. Small aircraft magnetos are also moulded from it. The range of applications of this plastic is increasing, as its strength and speedy moulding cycle commend it to manufacturers.

PHENOL ASBESTOS COMPOSITION

British Trade Name — Keebush.

American Trade Name — Haveg 41.

Characteristics — Strength: resistant to all acids, chlorine and some solvents: resistant to chemicals except those of an oxydising type: unaffected by rapid changes of temperature.

Available Forms — Pipes: cylindrical or rectangular shapes for constructional use.

Fabrication Methods — Moulding without hydraulic pressure.

General Applications — Entirely industrial, for equipment which must possess resistance to corrosion, such as acid storage tanks, pickling tanks, pipes and valves for food production factories. As hydraulic pressure is eliminated from the moulding technique, very large shapes can be economically produced, which are strong enough to be used in actual constructional work. The abrasion-resistance of this plastic is good, but even if scratches occur on the inside of an object made from it, no bad effects will follow, for the material resists corrosion throughout its entire thickness.

CAST PHENOLIC

British Trade Names — Bakelite Cast Phenolic: Catalin: Erinite.

American Trade Names — Bakelite Cast Resinoid: Baker Cast Resin: Catalin: Gemstone Marblette: Opalon: Prystal.

Characteristics — Translucency: transparency and opacity: very wide and brilliant colour range: high tensile and impact strengths: practically odourless: tasteless: non-absorbent: non-flam.

Available Forms — Cast polished sheets: cast unpolished sheets: clear and opaque casting resins: cast rods, tubes and blocks: two-tone castings: split mould shapes: profile castings: liquid cements: laminating resins: resin solutions in organic solvents.

Fabrication Methods — Casting: sawing: cutting: drilling: threading: carving: turning: milling: ashing: embossing: polishing.

General Applications — The gem-like appearance of this plastic has made it popular for consumer goods where this quality is 'of value, such as:

Beads	Costume jewellery
Billiard balls	Cutlery and fork handles
Book ends	Hairbrushes
Buttons	Handbag tops
Carved statuettes	Interior lighting fixtures
Chess and backgammon men	Jewel boxes
Cigarette boxes and cases	Portable radio housings
Clock cases	Safety razor housings
Cooking utensil handles	Shop display fittings

Sometimes the plastic is used to simulate a precious material such as onyx or tortoiseshell, but this imitative use is diminishing with a material so intrinsically beautiful. It can be opaque, translucent or transparent. There are also two-tone castings where two shades of the same colour merge almost imperceptibly into each other—first one shade predominating and then the other as the light strikes the object. In addition, there is a special confetti type in which small, multi-coloured chips are added to the transparent resin before casting. With such decorative possibilities, the imitative types are dull and poor in comparison.

But this is only one side of the picture, and cast phenolics have also a strictly utilitarian part to play in which they are as successful as in their decorative function. Petroleum pump gears made of cast phenolic are strong, almost silent in action, and resistant to oils and acids. Milk capping machines use revolving cap containers cast from this plastic, and housings for electrical suction and compression pumps for laboratory use, are also made from it. Other forms are used for adhesives and veneers, for impregnating resins and for laminating varnishes for translucent panels. (A note on these uses will be found on page 98).

Another development in the use of cast phenolics is concerned with a new type of very fluid low viscosity resin, which can be maintained in its fluid state over a period of months. This fluidity gives castings which show the finest detail and most minute features of the original pattern. The casting moulds used for this resin are usually of plaster, but wood, lead, nickel, brass, bronze, copper or tin can all be used. Casting is carried out at atmospheric pressure, and at comparatively low temperatures; when the cure is complete the usual machining operations can be employed if desired.

The importance of this new type of cast phenolic lies in the uses to which it can be and is being put. Originally produced for making moulds in which to cast certain oil drilling tools, it is now replacing large quantities of metal dies in aircraft construction. Aircraft machine tools such as hydropress form blocks, stretch press dies, assembly and drill jigs, and check fixtures, are all made from it. Assembly fixtures and dies for forming polymethyl methacrylate are also made from it, as well as dies for forming laminated phenolics. All these applications are not only economical from a financial aspect, but have greatly decreased the tooling time programme where they have been used—in some applications by as much as three months.

See Section III, plates v, vi, xxxvii, xxxviii, xli and xlviii.

UREA.

British Trade Names — Beetle: Mouldrite U: Scarab.

American Trade Names — Bakelite Urea: Beetle: Plaskon.

Characteristics — Rigidity: hard surface: odourless: tasteless: light in weight: dielectric strength: electrically non-tracking: transparent, translucent or opaque: first-rate light diffuser: unlimited colour range.

Available Forms — Powders for compression moulding: resin in solution for lamination, adhesives and textile treatment: heat-setting solution: cold-setting cement.

Fabrication Methods — Compression moulding: laminating.

General Applications — These are many and varied in type, and include a long list of consumer goods of which the following examples indicate the type:

Bathroom beakers	Fuse box covers
Bottle closures	Handwheels for water-softening
Buckles	plant
Butchers' trays	Insulation plates
Buttons	Light-switch cases
Clock cases and faces	Sandwich boxes
Cocktail shakers	Shields for safety sockets
Cosmetic containers	Shop display stands
Cruet sets	Radio housings
Door furniture	Table-ware (plates, cups and
Draining boards for sinks	saucers)
Freezing drawers and lids for	Transparent display boxes
refrigerators	Weighing scale housings
Funnels	

As this plastic does not discolour with age, and its colours have very little tendency to fade, even the palest pastel shades and clear water white may be used with safety.

Urea is also in general use for electric light fittings: table lamps, bowl reflectors, shields and shades. Here, its excellent light-diffusing quality, together with the pastel colours available, give fittings which are useful, decorative and reasonable in cost. Other typical uses include illuminated instrument dials, various electrical wiring accessories, control knobs and handles, and electrical controls.

Urea resins in solution are used for surface treatment of textiles and bonding agents for various kinds of lamination. (*See* bonding agents, page 98). Resin in solution is also used for giving high wet strength to paper which is used for food packaging and towels.

Some recent experiments in enclosing natural flowers in blocks of urea before it is set, have had interesting results. The flowers, after a year, still retained their freshness and original brilliance of colour; first conceived as a decorative item, this use opens up possibilities for preserving in their original state rare botanical specimens for study. Flowers have been preserved in transparent blocks of other plastics; urea, however, is the only one in which they retain their natural colouring. In others they keep their shape, but fade to a lifeless brown.

See Section III, plate xxvii.

MELAMINE

British Trade Name — Beetle Melamine.

American Trade Names — Melmac: Catalin Melamine: Plaskon Melamine.

Characteristics — Inert: odourless: tasteless: greater resistance to heat and abrasion than urea: unaffected by hot water, alkalies, organic solvents and weak acids: arc-resistant: great dielectric strength: available (in the alpha-cellulose filled type) in black, white and a good range of fast colours.

Available Forms — Moulding powder and granules (with alpha-cellulose filler): granular moulding powder (with mineral filler): moulding compound (with cotton rag filler): solid resins: resin in solution for laminating and adhesive use: resin for stoving enamel.

Fabrication Methods—Compression and transfer moulding: laminating.

General Applications — Melamine has been used in the manufacture of stoving enamels and finishes for a considerable time; its use as a moulding compound is a recent development. Compared with urea, its chief advantages are lower water absorption and greater heat resistance. The different types of filler employed give it different characteristics suitable for various purposes. When alpha-cellulose is used, the plastic resists boiling water, fruit juices and staining, and can be fabricated into lightweight and extremely durable table crockery. It is in considerable use in the U.S.A. for air line tableware, and the U.S. Navy also uses it for this purpose. Its durability (it can be dropped with impunity) and its lightness of weight make it ideal for both ship and aeroplane crockery. Plastic crockery is attractive and hygienic, and its resistance to staining and acids has always been an asset: the alpha-cellulose melamine's resistance to boiling water suggests a use for such table-ware, not only in transport services, but in the home.

Melamine can also be used for button-making, costume jewellery, and other decorative trifles; but it is probably in more utilitarian applications that it may find its widest use. Here, its high dielectric strength and arc- and heat-resistance indicate its use for various electrical parts. In the electrical industry the mineral filled type is used for distributor heads, terminal blocks, and also in the ignition systems of tanks, aircraft and tractors. The experience gained in moulding this plastic for war material is certain to suggest many future applications. A note on the uses of melamine for stoving enamel will be found in the section on various plastic finishes, on page 101. A melamine solution is also successfully used for impregnating packaging paper to give it high wet strength.

See Section III, plate XXVIII.

THERMO-PLASTIC GROUP

CELLULOSE ACETATE

British Trade Names — Armourbex: Bexoid: Celastoid: Cellastine: Cellomold: Cinemoid: Clarifoil: Dorcasite: Doverite: Erinofort: Isoflex: Novellon: Wireweld.

American Trade Names — Bakelite Cellulose Acetate: Fibestos: Lumapane: Lumarith: Nixonite: Plastacele: Tenite I: Vimlite: Vuelite.

Characteristics — High dielectrical and mechanical strengths: resistant to salt water and weak acids: opaque or transparent: light in weight: easy to fabricate: resistant to mechanical abrasion: wide and brilliant colour range: non-flam.

Available Forms — Powders for compression, injection and extrusion moulding (granules or pellets): tubes: rods: profile rods: rigid or flexible sheets in standard sizes: films and foils in continuous lengths.

Fabrication Methods — Compression, injection and extrusion moulding: swaging: deep drawing: forming: blowing: turning: punching: sawing: drilling. Can be polished by the buffing wheel or tumbler barrel methods. Can be metal-plated, metal-inlaid, lacquered and die-stamped. In its flexible sheet and film forms, it can be printed, embossed and laminated to cardboard, paper, aluminium and lead foil, and thin textiles such as scrim. It can be cemented with special cements.

General Applications — A beautiful and versatile plastic, its immense diversity of applications, decorative and utilitarian, includes:

Architectural profile mouldings and cut strip for cornices, pelmets, skirting boards, panel beading, etc.	Combs
	Cosmetic containers
	Costume jewellery
	Curtain rings and hooks
Artificial flowers	Cutlery handles
Blotters	Door furniture
Buckles	Electric light fittings
Buttons	Eyeshields
Calendar frames	Fish baits
Cigarette cases and boxes	Fishing-rod parts
Clock cases	Fountain pens
Clothes hangers	Furniture trim and edgings

Games counters	Table barometer and thermo-
Goggles	meter cases
Hairdrying machine shields	Telephone housings and micro-
Hairpins, slides and curlers	phones
Handbag tops and closures	Toilet brush backs and handles
Kitchen accessories	Toothbrush handles
Luggage handles	Toys
Mechanical pencils	Transparent containers
Paper knives	Typewriter and adding machine
Safety razor housings	keys
Shoelace tips	Umbrella handles and spoke tips
Shop display fittings	Vanity cases
Spectacle frames	

Cellulose acetate is also used for aeroplane cockpit enclosures, bomber noses, machinery guard shields, all sorts of tool handles, blast-proof window glazing (with an inner sandwich layer of wire or string mesh), horticultural cloches, cartridge cases for flares, oil cans, oil tank floating gauges, cistern float balls, instrument panels, directional signal lenses, first-aid kit boxes, gunstock sights, binocular parts, and many other objects of a similar type. Wire, coated with cellulose acetate, is used for both electrical insulation and as a decorative material.

The deep drawing of cellulose acetate sheet has been greatly developed, and this method is used to produce troughs and shields for fluorescent lighting. Deep drawing is also used to make all sorts of transparent advertising and display containers; large objects of this kind, from simple to fairly intricate shapes, may thus be quickly and economically made. Metal-plated sheet, plain or embossed, is used for lamp-shades, handbags, fancy goods and display purposes. In ordinary light this material seems opaque, but when lit from the back its translucency is revealed.

Another use for cellulose acetate is to wind extruded strip at tension over a revolving mandrel; the result is a tube, of which the component strip can be pulled apart. When tension is removed, the separated strip springs back into position. These tubes can be mounted over metal or wooden cores and used for table legs, banister rails and similar purposes; highly decorative effects can be obtained by the use of white or coloured transparent tubes over metal cores; but the tubes enjoy their greatest use without an additional core, as casings for fluorescent lighting.

The sheet, foil and film forms are used in all kinds of packaging, from wrappings of flexible sheet to "windows" in cardboard boxes, which allow the contents to be seen. The sheet can be printed and embossed; laminated to cardboard, paper or aluminium foil it is used for grocery packages, grocery and sweet bags, gift wrappings and so forth.

A special type of this plastic excludes those rays which cause sunburn—this type is used for cockpit enclosures for high altitude aeroplanes.

See Section III, plates IV, XIII, XIV, XVIII, XXII, XXIV, XXV, XXXI, XXXIV, XXXV, XXXVIII, XXXIX, XL, XLI, XLIII, XLVI and XLVII.

CELLULOSE ACETATE BUTYRATE

British Trade Names — None.

American Trade Names — Rexenite: Rextrude: Tenite II.

Characteristics — Dimensional stability: tensile strength: low moisture absorption: transparent or opaque: wide permanent colour range: non-flam.

Available Forms — Powders for compression, injection and extrusion moulding (granules or pellets): rods: tubes: profile shapes: extruded strip: extruded filament: films.

Fabrication Methods — Compression, injection and extrusion moulding: machining.

General Applications — As a general rule, it may be assumed that cellulose acetate butyrate is suitable for any purpose for which cellulose acetate can be used, but that there are other applications for which its greater strength is particularly suitable. It has a higher degree of water-resistance than cellulose acetate and better insulating qualities, while the film forms are highly elastic. There are also uses for extruded cellulose acetate butyrate filament which are not possible with cellulose acetate.

Among the special uses for this plastic, is a rattan-like fabric made from basket-woven strip which is used for the upholstery of passenger transport, and for garden and porch furniture. Cellulose acetate strip can be made into the same material, but it is inclined to split in wear, whereas the cellulose acetate butyrate strip will stand up to hard and prolonged wear. As the plastic is impervious to staining, anything spilt on it can be wiped off with a damp cloth; this quality together with the great variety of colours available, both in the opaque and transparent types, gives it a large field of utilitarian and decorative uses. Thinner gauges of the extruded strip can also be power- or hand-woven into a lustrous and sparkling material much used for handbags, belts, ladies' hats, braids, braces and lampshades.

Extruded cellulose acetate butyrate filaments and small diameter tubes can be woven or braided into a lace-like fabric which is in great demand for dressmaking and millinery trimmings, for collar and cuff sets, and braids from which evening shoe uppers and whole hats can be made. This material can be stiffened by steaming, and needs no sizing; it will not crush or soil in wear, and has a permanent lustre and colour.

Typical moulding applications of cellulose acetate butyrate are for garden and other tool handles, pistol grips, industrial oil cups and bottles, hose couplings, window blinds for 'buses, fishnet floats, army whistles, army bugles, transparent piping coils for the brewing trade, transparent cases for lightning arrestors, vacuum venom extractors, and lightweight housings for rotary pneumatic drills which enable women operatives to use the drills with great accuracy and the minimum of fatigue. An agricultural development in the use of this plastic is the extruded one-inch tubes which are successfully used for farm irrigation. After extrusion, these tubes are bent to an angle of 120°, and used to syphon water from the irrigation ditch to the growing plants.

Among applications of the profile shapes, is a piano type of hinge, which weighs considerably less than a metal one, but is suitable for relatively heavy construction. It can be assembled either with adhesives, screws or rivets, is rust-proof and dirt-resisting, and will endure long, hard wear without denting, chipping or cracking. It is used for aeroplane tool boxes, oil immersion boxes and radio, map and chart cases.

See Section III, plates II, III, XV, XVII, XXVI, XXXI and XLV.

CELLULOSE NITRATE

British Trade Names — Xylonite.

American Trade Names — Amerite: Celluloid: Hercules Cellulose Nitrate Flake: Nitron: Nixonoid: Pyralin.

Characteristics — Tough: water-resistant: easy to fabricate and cement: opaque, translucent and transparent: good colour range: inflammable.

Available Forms — Sheets: films: rods: tubes: beading: lacquers: emulsions.

Fabrication Methods — Blowing: extrusion and compression moulding: cutting: drawing: drilling: punching: sawing: turning: deep drawing: polishing. The sheets can be printed and embossed.

General Applications — This plastic is the oldest of those in commercial production, and is used for many purposes. Its earliest uses were chiefly decorative, and for small consumer goods; modern methods have much extended its range of applications. Its inflammability is its chief drawback: to some extent this may be overcome by the use of fire-retardant plasticizers in its manufacture. Its colours are excellent, and it can be produced in a variety of mottled, marbled and multi-coloured effects.

The blowing method of fabrication makes possible the production of light, brightly coloured and inexpensive toys of many different kinds, which will float. Table tennis balls annually consume a huge volume of this plastic, and another very large use is for cinema film. Floating cistern valves, dominoes, piano keys, drum and accordion coverings, spectacle frames, covers for wooden shoe heels, buckles and novelty jewellery, combs, fountain pens, tooth-brush handles, tool handles, cutlery handles, containers for soldiers' kit, range-finder parts, labels, eyeshades, are all made from cellulose nitrate, and illustrate a few of its diverse uses.

(A note about the uses of cellulose nitrate as a coating is included in the section on plastic coatings on page 101).

ETHYL CELLULOSE

British Trade Name — B.X. Ethyl Cellulose.

American Trade Names — Ethocel: Ethocel P.G.: Ethofoil: Hercules Ethyl Cellulose Flake: Lumarith E.C.

Characteristics — Low density: great strength: dimensional stability: low-temperature impact resistance: low water absorption: alkali-resistance: excellent electrical properties: light weight: odourless: easy to fabricate.

Available Forms — Powders for compression, injection and extrusion moulding: extruded rods, tubes and profile rods: sheets: films and foils: flakes for formulation to coatings, lacquers and emulsions.

Fabrication Methods — Compression, injection and extrusion moulding: swaging: deep drawing: cutting: sawing: punching: drilling: turning: polishing.

General Applications — An American manufacturer of this plastic calls it a "tough two-fisted plastic"—an apt description. While its use is, as yet, more limited than other plastics of the cellulose family, it possesses two important properties. These are its extreme toughness, and its ability to stand up to shock conditions under varying temperatures. Its colour range is not as good as that of the other cellulose plastics; research is in progress on the improvement of its colours, meanwhile it is advisable to avoid pastel shades, as they are apt to be slightly tinged with yellow. Where real ability to take hard knocks is required of a plastic, ethyl cellulose will give exceptional results.

For moulded applications, its low temperature resistance makes it useful for many refrigerator parts such as drip trays, freezing unit doors, sealing frames and ice cube freezing trays and cups. The American Army has replaced its metal field canteens by a canteen moulded from ethyl cellulose, which is serviceable in any climate, light in weight, and can be freely handled when filled with either hot or cold liquids, the plastic acting as an insulator. This canteen will stand up to the roughest use—in tests it has even withstood a ten-foot drop on to a concrete floor. Other American Army uses for this plastic are coston tubes to house dry batteries, aircraft control knobs, cockpit de-frosting nozzles for aeroplanes (here it has replaced brass), and extruded aircraft window frames and table edging trim for barrack furniture. It is also used for electric light bulbs for life rafts, and its good electrical properties make it suitable for microphone handles. The U.S. Army Signal Corps uses a communication cable with an ethyl cellulose core which remains flexible and tough throughout the temperature range to which the cable may be subjected.

Besides the moulding type, there is ethyl cellulose sheet, which is a clear, transparent material with great flexibility, ductility and toughness; it is easily fabricated into all kinds of rigid or flexible containers which are very suitable for packaging a wide range of consumer goods; small drawn covers of this material are used to protect 20 mm. anti-aircraft guns from dust, dirt and salt spray. As the sheet is entirely odourless and non-toxic, it is particularly suitable for the packaging of foodstuffs, in which capacity it affords complete protection from dirt and humidity. The sheet can be printed in any desired colour, and the ease with which it can be formed makes it a valuable packaging material.

Ethyl cellulose is also used in coatings and lacquers. (A note on these applications appears on page 103).

See Section III, plates xxvi and xxvii.

POLYMETHYL METHACRYLATE

British Trade Names — Diakon: Kallodent: Perspex: Transpex I.

American Trade Names — Acryloid: Crystalite: Lucite: Plexiglas.

Characteristics — Great transparency and light transmission: strength: rigidity: water and weather resistance: dimensional stability: dielectric strength: light in weight: unlimited colour range: non-flam.

Available Forms — Powders for compression and injection moulding: cast sheets: moulded sheets: rods: tubes: granular solids: liquid monomer.

Fabrication Methods — Compression and injection moulding: casting: forming: carving: sawing: drilling: threading and tapping: routing and shaping: polishing.

General Applications — One of the most beautiful of all plastics, polymethyl methacrylate has a versatile range of decorative and utilitarian functions. Its extreme clarity and power to transmit or "pipe" light round a curve or angle, make it an ideal material for lighting fixtures. Table lamps and floor standards, ceiling and wall fixtures are all made of it, and its "edge lighting" powers allow the actual source of light to be hidden, so the whole fixture seems to glow from within itself. Entire ceilings have also been made of cast blocks which, when lit from the back, illumine a room with an agreeably diffused light. This power of light transmission is also exploited for all sorts of directional signs, house name and number plates, and for shop display fittings. Some of the most important of its uses are in aircraft construction, where it is used for:

Bomber noses	Housings for radio and range-
Cabin and cockpit enclosures	finder antennæ
Escape doors	Landing light covers
Fuselage panels	Portholes and windows
Gun turrets	Windshields, etc.

Its advantages for these purposes are its clarity and lightness of weight and its weather resistance. While not actually unbreakable, it can withstand a hard direct blow; when held rigid a blow hard enough to break it will produce splinters, but mostly in the form of large, dull-edged pieces.

Other typical applications are:

Bathroom fittings	Door furniture
Bedheads and footboards	Hair, clothes and nail brushes
Bowls and vases	Jewellery
Cake and pastry knives (both blades and handles)	Machine guards
	Machine inspection windows
Cigarette boxes and cases	Safety goggle lenses and eye-
Contact lenses	shields
Cosmetic containers	Salad spoons and forks and vege-
Cutlery handles	table servers
Decorative door lintels and window pelmets	Switch covers
	Tables, chairs and stools
Dentures and false teeth	

and many other similar uses too numerous to list in detail.

There is a heat-resisting grade of this plastic which will stand up to eight minutes' immersion in boiling water, or eight minutes in a dry heat oven at a temperature of 212°F. This grade is used for surgical and dental instruments, which can be illuminated from a battery in the handle, allowing the surgeon to light up remote cavities in a way never possible before. Curved throat inspection instruments and curved dental lights are two examples of such applications. Other uses for this heat-resisting type are:

Aircraft instrument lenses
Airport signal lenses
Automatic feeder and lubricator parts
Cases for lamps and batteries
Compressed air line lubricators
Dial and meter faces
Flying light lenses
Insulation for radio equipment
Railway signal light lenses
Transparent covers for industrial lights

Polymethyl methacrylate is also used as a coating material. (A note on this use appears on page 102).

See Section III, plates XIV, XVI, XIX, XXI, XXIII, XXXVII, XXXVIII, XLIV and XLVII.

POLYSTYRENE

British Trade Name — Distrene.

American Trade Names — Bakelite Polystyrene: Loalin: Lustron: Styraloy: Styron.

Characteristics — Low specific gravity: water absorption almost nil: dimensional stability: good low-temperature strength: ability to transmit light: high refractive index: outstanding high-frequency insulating properties: resists chemical action: unlimited colour range.

Available Forms — Powders for compression, injection and extrusion moulding: sheets: film: rods: tubes: sections: cast sheets.

Fabrication Methods — Injection and extrusion moulding. (Polystyrene can also be moulded by compression, but this method is seldom used). Machining.

General Applications — Its resistance to chemicals, in particular to acids, has led to its use for medical bottle closures and containers, as well as parts for blood analysis kit. Entire bottles are also made from it for such chemicals as hydrofluoric acid. Its exceptional dimensional stability makes it suitable for mouldings where absolute precision and accuracy are needed. Its low temperature strength and negligible water absorption and complete freedom from odour have led to its extensive use for many interior refrigerator parts.

Incidentally, the largest single unit ever moulded by the injection process is a complete polystyrene sealing frame for a refrigerator freezing unit. Its high dielectric strength recommends it for special uses in high-frequency applications. Storage battery containers, covers and functional parts in radio sets, condenser housings and coil forms for radio and television sets are all made from it. Polystyrene films are suitable for condenser foil or cable wrapping, and solutions made from it can be used to form insulating lacquers for radio coils.

Its clarity and ability to transmit light (in the same way as polymethyl methacrylate, around curves or angles) are utilised for all types of electric light fittings. Moulded tiles are made from it, which may be fixed together to form flush ceiling lights or backing for show cases and other display and exhibition uses. These tiles can be either permanently or temporarily joined together, and are available in many colours. Polystyrene is also a very popular material for fruit, sweet and salad bowls and dishes, portable radio housings, advertising displays, jewel and cigarette boxes, control knobs, combs, buttons, jewellery, cosmetic containers and many other forms of consumer goods, where its attractive colours and crystal clarity give a jewel-like and luxurious effect at a most reasonable cost.

Another type of this plastic has been produced which is more flexible than the ordinary polystyrene; this is flexible down to as low as $-70°F$. and is sufficiently strong to protect communication cables which have to be dragged across hot sand or frozen ground. This type is also used to insulate lead-in wires on bombers, and as flexible antennæ spacer insulation for aircraft, and as insulation for ignition systems for aircraft where standard type materials fail because of high voltage corona breakdown.

See Section III, plates xx, xxiv, xxxviii, xl, xlii and xlvii.

POLYVINYL CHLORIDE

British Trade Names — B.X. PVC: Chlorovene: Welvic.

American Trade Names — Koroseal: Vinylite.

Characteristics — Flexible and rubber-like, or rigid, according to grade: flexible type has considerable elongation and good recovery: high tensile and tearing strengths: inert to oxidation and weathering with freedom from cracking in use: negligible water absorption: good electrical properties: non-cracking at temperatures down to $-30°C$.: resistant to most corrosive liquids and inert to most organic solvents: great resistance to abrasion: non-flam: low specific gravity.

Available Forms — Flexible sheet in continuous lengths: rigid transparent sheet: flexible and rigid tubing: powders for extrusion moulding: extruded shapes and sections: sheet laminated with textiles.

Fabrication Methods — Extrusion and injection moulding: polishing: laminating: embossing: machining: drawing: blowing: rolling: stamping: hot sealing.

General Applications — This is a most useful plastic. It has two grades, flexible or rigid, which give it considerable scope. The flexible sheet is used for articles of clothing such as boot and shoe soles, heels and uppers, belts, suspenders, braces, waterproof aprons and raincoats; it is also made into ladies' handbags, note-cases, tobacco pouches, sponge-bags, and artificial flowers for millinery. Its water-resistance makes it an attractive material for shower, bathroom and kitchen curtains. Utilitarian uses for the sheet include such items as gaskets, electric cable insulation, aircraft sealing and printed adhesive tape. The sheet is available in any colour, including pastel shades and its own natural tone. It can be applied to a variety of fabrics, and with this backing makes serviceable and good-looking upholstery material for all types of transport and domestic furniture. Unbacked, the thick, flexible sheet in its natural colour is also used as a transparent protective cover for delicate silk or brocade upholstery; thick, flexible sheet is also cut into strips and basket-woven into chair and settee seats. It has great resilience, so that no other form of springing is necessary when the basket-woven strip is used; it is exceptionally hard-wearing and easily cleaned.

The thin flexible sheet is also laminated to heavy canvas and from this, gaskets, instrument panels and account book covers are made.

The rigid, transparent sheeting is used for aircraft enclosures, calculating, drafting and navigating instruments, engraved labels, instrument and radio dials, recording charts, storage instruments, and similar uses where non-shrinking and dimensional stability are required.

The flexible and rigid tubings are available in all colours and have a number of industrial and electrical uses. Among these are pipes for plumbing purposes, or for chemical plants to meet corrosive conditions; it is also used where an inert material is needed for food products such as milk or beer or similar liquids, or to pipe corrosive fluids, or petrol. The advantage of this use of transparent tubing is that it allows the flow of liquid to be observed at all times.

The flexible type of tubing is also used for insulating sleeving for radio and telephone work; thin tubes of the flexible type are made into braids and webbing for millinery, collar and cuff sets, glove edging, evening shoe uppers and other kindred uses.

The extrusion compound has a standard range of twelve colours and its natural tint; it is prepared specially to meet standard electrical specifications, and the tube made from it is used for all types of cable dielectric and sheathing insulation. It effectively replaces rubber for domestic, aircraft, radio and telephone cables and insulating sleeving. Extruded polyvinyl chloride is also supplied in shapes and sections which have various uses such as curtain rails, towel racks, architectural mouldings, and furniture trim.

See Section III, plates i and xxxi.

POLYVINYL ACETALS

British Trade Names — Bexone: Bexone F.

American Trade Names — Alvar: Butacite: Butvar: Formvar: Saflex: Vinylite X.

Characteristics — Great clarity, toughness and adhesiveness: resistant to moisture: stable to light and heat.

Available Forms — Flexible sheets in continuous lengths: rigid sheets: moulding compounds.

Fabrication Methods — Injection and compression moulding: for safety glass, heat and pressure are applied in bonding the plastic sheets between sheets of glass.

General Applications — Used largely for the manufacture of some types of safety glass for which its characteristics are particularly suitable. Occasionally the plastic interlayer is allowed to protrude beyond the glass surface sheets, thus affording a convenient projecting edge for attaching the glass to the window frame. This type of plastic is also used extensively for adhesive laminations and the coating of fabric. Some forms have been successfully developed and moulded to replace rubber in certain electrical and other applications.

POLYVINYLIDENE CHLORIDE

British Trade Names — None.

American Trade Names — Saran: Vec: Velon.

Characteristics — Great tensile strength: great resistance to fatigue and abrasion: resistant to water and chemicals: high dielectric strength: non-inflammable: rigid or flexible according to grade.

This plastic was first commercially used in 1940. Though "Saran" is the trade name used by the original makers, they have released it for use as a generic name to describe any polyvinylidene chloride plastic. Its most important attributes are great strength and phenomenal resistance to chemicals—acids, alkalies, salt solutions and organic solvents. It is also immune to attack by bacteria and fungi.

Available Forms — Powders for compression, injection and extrusion moulding: extruded strip and tubes: extruded filaments: transparent film.

Fabrication Methods — Compression, injection and extrusion moulding: and, for the extruded filaments, power- and hand-loom weaving, braiding and knitting.

General Applications — The light, tough and flexible tubing, in its natural yellowish colour, is in extensive use in chemical laboratories; it is also used for instrument lines on many types of machine where short lines are necessary to pipe water, oil or petrol. It is also successfully used in plumbing systems, particularly where the water is hard and corrosive. It is easier to handle than copper, and the other standard types of tubing. It will withstand freezing, is non-scaling, will accommodate high and low pressures, and heat up to 175°F., and is non-sealing. It can be supplied in a variety of sizes and thicknesses.

Rigid pipes are also extruded from this plastic in sizes to match standard iron pipes; the plastic pipes are only one quarter as heavy as the same size iron pipe, and have great strength. The rigid type of polyvinylidene chloride has the same attributes as the flexible tubing: chemical resistance, non-scaling, resistance to freezing, and accommodation to high and low pressures. These pipes can be welded, threaded, heated and bent, and are accompanied by injection-moulded fittings of the same plastic. The fittings include coupling nuts, union couplings, half-union couplings, ties and tube-pipe elbows. Other moulded applications of polyvinylidene chloride are chiefly those demanding dimensional stability and chemical inertness, and include such items as acid equipment, machine parts, special pistons, atomiser parts, carburettor floats, plating bath equipment and paint-brush and air-gun handles.

The film of this plastic is an exceedingly tough and flexible fabric with good transparency and great water and chemical resistance. It can be used for all sorts of packaging from food products to vital metal parts, tools and mechanical equipment. It retains flexibility at extremely low temperatures, forms a complete barrier to vapour and moisture, has great tensile strength, is incombustible, and withstands the "ballooning" effects of high altitudes; the latter quality makes it important for the packaging of air-borne goods.

Perhaps the most interesting uses of polyvinylidene chloride are those to which the extruded filaments are put; these are described in the section on plastic filaments, on page 94.

POLYTHENE

British Trade Name — Alkathene.

American Trade Name — Poly-Ethylene.

Characteristics — Translucent: flexible: great resistance to water and chemicals: low density: special electrical properties such as very low power factor and dielectric constant, high resistance powers and dielectric strength: toughness: good mechanical properties: easily fabricated and coloured.

Available Forms — Chips: strips: rods: films: powders for extrusion, compression and injection moulding. (It is supplied in various grades according to the purpose for which it is intended).

Fabrication Methods — Extrusion, compression, injection and impression moulding: machining: welding.

General Applications — These are greatest in the electrical field, where its combination of mechanical and electrical properties make it an excellent material for cables and accessories for high-frequency and high-voltage work, such as solid insulated and air-spaced high-frequency cables. Its great resistance to water makes it possible to dispense with the lead sheath usually employed to protect power cables. Moulded electrical parts made from polythene include cable ends, high voltage bushings and condenser dielectrics. Its great chemical resistance commends it for various applications in chemical plants. For instance, bottles moulded from polythene can be used as containers for fluoride and hydrogen fluoride: though the surface is liable to attack, such bottles can remain in effective use over a very considerable period.

Polythene may also be fabricated in the same way as the other thermo-plastics to produce all sorts of fancy goods and moulded containers. (A new technique known as "impression moulding" may be used: it is described in the section on Fabrication Methods, page 105). This plastic has a natural, translucent, milky-white appearance; or it is sometimes slightly grey or pink, without the addition of colouring matter. It may be readily coloured to any desired shade: in darker colours it becomes opaque, but in paler

shades it retains some of the original milky character of its un-coloured state, which gives the colours a soft, attractive quality.
See Section III, plate xi.

NYLON (Moulding Type).

British Trade Name — Nylon.

American Trade Name — Nylon FM.1.

Characteristics — Extreme toughness: flexibility: high temperature resistance: outstandingly low density: low water absorption: tensile strength: dielectric strength: slow burning: resistant to esters, ketones, alkalies, common solvents and weak acids: not resistant to phenol, formic acid and concentrate mineral acids.

Available Forms — Powders for extrusion, compression and injection moulding.

Fabrication Methods — Extrusion, compression and injection moulding. (Injection moulding is the most often used).

General Applications — Hitherto, nylon has been entirely used in its yarn or monofilament forms.* Its use as a moulding material is recent, and is still, to some extent, in the experimental stage. But it has proved successful when used for many heat-resisting parts. These applications form its major outlet, and here its extreme toughness and high temperature softening point permit its use in ways impossible with other thermo-plastics. Its high service temperature under load (approximately 275°F.) much exceeds that of other moulded thermo-plastics: it can also be injected around thin sections, and has the capacity of flowing around complicated inserts. Typical applications are a spool used for winding coils for actuating aircraft instruments, and a switchette housing: both these objects must resist operating temperatures capable of distorting other thermo-plastics, and must also have unusual toughness and flexibility; such requirements are fully met by the nylon moulding powder.

Another use is a moulded slide fastener which is immune to the action of dry-cleaning solvents, and undistorted by the temperature of ironing. The nylon forms a strong bond with the fibres of the cloth tape at the time of injection, thus eliminating the need for cementing the slide to the tape, which is usual in the manufacture of other plastic slide fasteners.

The basic colour of nylon is a light, translucent amber. The lack of effective solvents for this type of plastic makes the incorporation of dyes into it a difficult problem, and also complicates the development of a satisfactory cementing technique. So far, this plastic is

* *See* description of these uses in the section on plastic filaments, page 93.

important more for its mechanical properties than because of its appearance, but these properties are of first-rate value.

See Section III, plate XII.

EXPANDED PLASTICS

Plastics of this type can be more conveniently considered as forming a group of their own, though they may be either thermo-setting or thermo-plastic according to the resin chosen for expansion. They are the result of many years of intensive research, and form an entirely new series of light density materials. The term "expanded plastics" is used to identify a type of cellular structure which is independent or non-intercommunicating, unlike those of a spongy or porous nature. This cellular structure provides materials which are permanently buoyant and have an almost negligible water absorption, very low density, high strength to weight ratio, and exceptionally low thermal conductivity. Their properties can be varied to suit special requirements, by choosing specific plastic composition and density which can be controlled as desired.

THERMO-SETTING EXPANDED PLASTIC

British Trade Name — Thermazote.
American Trade Names — None.
Characteristics — Non-inflammable: odourless: water-resistant: buoyant: capable of resisting temperatures as high as 250° to 300°C.
Available Forms — Sheets.
Fabrication Methods — Cutting: sawing: drilling: veneering.

THERMO-PLASTIC EXPANDED PLASTIC

British Trade Name — Plastazote.
American Trade Names — None.
Characteristics — Exceptionally high impact strength: exceptionally high modulus of elasticity under compression: low thermal conductivity: water resistant: buoyant.
Available Forms — Sheets.
Fabrication Methods — Cutting: sawing: drilling: shaping: veneering.
General Applications of thermo-setting and thermo-plastic expanded plastics — The thermo-setting type is manufactured in densities between 7 and 30 lbs. per cubic foot. At its lowest density it has a compression strength of approximately 100 lbs. per sq. inch, and

at this density 1 cubic ft. of the material will support 55 lbs. of soft iron in fresh water. It is particularly useful for thermal insulation, and for constructional purposes, especially when veneered with plywood or lightweight metal alloys.

The thermo-plastic type has a high compression strength of 170 lbs. per sq. inch at the standard density of 6 lbs. per cubic ft. Its thermal conductivity at this density is 0.24 B.T.U., and its modulus of elasticity under compression is about 10,000 lbs. per sq. inch. It is in extensive use as a low density medium for sandwich construction, and as it can be readily shaped into curved panels, it eliminates carpentry and joinery when used for aircraft construction.

These are some of the uses of expanded plastics:

Buoys, floats, pontoons and life-saving apparatus. Insulating material for the refrigeration industry, for meat, fruit, ice-cream vans and containers fixed to all movable cold storage chambers, coolers and finishing rooms. General ship insulation, domestic refrigeration, pathological containers, insulation of bulkheads on cars, 'buses and ship's cabins, including deck housings of boats and yachts. Insulation of low-temperature scientific apparatus. In aircraft construction: struts, wireless masts and tail fairings, veneered panelling for flooring and partition walls and instrument panels.

PROTEIN PLASTIC GROUP

The oldest plastic in this group—casein—is the only one which is produced in large commercial quantities. Much experimental work has been done on the production of plastics with various protein contents, such as cotton seed, green coffee berries, scrap leather, soya beans and so forth. With the exception of the soya bean and coffee berry plastics, all these are still in the early stages of experiment. A coffee berry plastic under the name of "Caffelite" is in commercial production in America; soya bean plastic has reached an advanced experimental stage. Work on this has been carried out by various organisations, including the Bureau of Agricultural and Industrial Chomistry of the U.S. Department of Agriculture. The Bureau has produced a low-cost soya bean moulding compound in combination both with phenol formaldehyde and urea formaldehyde: the bean has been found to be incompatible with methyl methacrylate, vinyl or cellulose acetate, so that the plastic produced is of the thermo-setting type, though it is said to possess some degree of affinity with the thermo-plastics.

Another form of protein—keratin—a substance derived from hooves, horn, hair and feathers—is in use in the U.S.A. as an extender for phenol formaldehyde plastics. This use was brought about by the enormous war demand for these two chemicals, and their consequent high priority, and the need for an increased use of extenders. It is doubtful if this keratin-extended phenol has any particular peace-time value, though keratin is in limited commercial use in Great Britain as an extender for a urea plastic, chiefly used for button-making.

Little information is available as to other experimental work in protein plastics; but in view of the great developments and new uses for what may be termed the "classic" plastics, it is unlikely—apart from the soya bean plastic—that users of plastics need be seriously concerned with the experimental types. For all practical purposes, casein leads the field in its own particular group.

CASEIN

British Trade Names — Dorcasine: Erinoid: Keronyx: Lactoid.

American Trade Names — Ameroid: Galorn.

Characteristics — Easy to machine and polish: hygroscopic: excellent colour range: non-flam.

Available Forms — Standard size sheets in standard thicknesses (special thicknesses can be made to order but not thinner than 2 millimetres): rods: tubes: discs or button blanks: two- and three-colour laminated sheet for button-making (specially shaped pieces punched from sheet can also be supplied to order).

Fabrication Methods — Cutting or grinding of fully cured articles to a desired shape: chemical hardening of uncured objects by immersion in a solution of formaldehyde.

General Applications — The principal use is for buttons of all kinds, from plain to highly decorative; plain and fancy buckles are also made from it, as well as the bulk of the non-metal knitting needles on the market. Other uses are poker chips and games counters, beads, a little costume jewellery, fancy trimming accessories, push buttons, handbag handles and closures, umbrella handles and small aeroplane bushings. The colours in casein are of particular brilliance and wide range; mottles and marblings and similar effects are also possible with it; the two- and three-colour laminated sheet give the designer of small objects an attractive decorative material. The ease with which the plastic can be machined and polished, together with

the excellence of its colours, make it useful as a material from which
to produce small and inexpensive articles of considerable beauty
and luxurious effect.

PLASTIC LAMINATES

Plastic laminates are of two types: industrial and decorative.
For the industrial type, the impregnating resin is usually a phenol
formaldehyde; for the decorative, the impregnating agent is urea
formaldehyde; melamine formaldehyde is also used.

The manufacturing process for these laminates is the same for
both types and only differs in the choice of the impregnating medium
and of the reinforcing material used to build up the finished sheet.
In industrial laminates, the reinforcing medium is mostly a heavy
cotton textile such as duck, canvas or sheeting or kraft paper;
asbestos sheet is also used to produce a special fire-resistant type,
and glass fibre cloth and mats are also used to make thin laminate
sheets for special electrical purposes. The actual weight of the
reinforcing material is determined by the ultimate use of the finished
material. For decorative laminates, different grades of paper are
the most usual reinforcing material, though lightweight textiles are
also used occasionally.

The reinforcing sheets are impregnated by various methods
according to the desired amount of resin content. The methods
include immersion in a resin solution, coating on one side and knifing
off the excess and coating by means of coating rollers. Whatever
form the impregnation takes, the sheets are next dried to remove
excess solvents, and finally formed between steel plates in a hydraulic
press.

Industrial laminates are also made in the form of rods and tubes.
The tubes are of two kinds, rolled and moulded. The rolled tubes
are made by wrapping impregnated strips of filler material around
a metal mandrel and curing these in an oven, without the application
of pressure. Moulded tubes are also first wrapped around a mandrel,
but the final operation is that of curing in a mould under heat and
pressure. Rods are made by rolling strips of impregnated material
around a mandrel of very small diameter; this is removed before
the curing operation, which takes place in a mould under heat and
pressure. This final operation seals up the small cavity left by the
mandrel and forms a solid rod.

Industrial laminates can be moulded, and many kinds of industrial
machine parts, which need great impact strength, are made from
them.

British Trade Names — Bakelite: Bushboard: Celloron: Delaron: Fabroil A: Formapex: Formica: Hydulignum: Micarta: Panilax: Paxolin: Permali: Philitax: Pirtoid: Traffolyte: Trafforoll: Tufnol.

American Trade Names — Aqualite: Celoron: Dilecto: Duraloy: Formica: Insurok: Lamicoid: Micarta: Panelyte: Parkwood-Textolite: Phenolite: Ryertex: Spauldite: Synthane: Textolite.

INDUSTRIAL LAMINATES

Characteristics — Easy to machine: easy to mould: inert: insoluble: high dielectric strength: dimensional stability: favourable strength to weight ratio: low water absorption: resists corrosion: light weight: low thermal conductivity: low co-efficient of friction: low co-efficient of thermal expansion: uniform co-efficient of expansion: resilient: high impact strength.

Available Forms — Sheets: rods: tubes.

Fabrication Methods — Machining: moulding.

General Applications — Industrial laminates are dense, solid materials, with a smooth surface which may be highly polished, satin-finished or matt. The standard colours are black and brown. These materials have great strength, are less liable to fracture than certain metal alloys, and are free from variations in structure. Some typical applications are:

Aerial masts	Fairlead pulleys and guides
Bearings	Fuse panels
Camshaft drives	Gears (spur, helical, mitre, worm and herringbone)
Circuit breaker barriers	
Coil frames and spools	Parts for manual and automatic telephone systems
Conduits for electrical wiring	
Conveyor belt tracks and sprockets	Pickers for weaving looms
	Plating barrels
Electrical insulation sheets and parts	Radio valve holders and bushings
	Rayon producing equipment
Factory truck wheels	Volt transformers

See Section III, plates VII and VIII.

DECORATIVE LAMINATES

Characteristics — Extreme hardness and durability: non-inflammable: shatter-proof: resists abrasion: non-warping: non-shrinking: resistant to fruit acids, alcohol and other solvents, and mild alkalies: resistant (in some grades) to cigarette burns: translucent in some grades: low water absorption: large colour range.

Available Forms — Veneer sheets: sheets with compressed fibreboard cores: sheets with steel cores: a few moulded shapes for skirting boards and window sills.

Fabrication Methods — Machining: cold bending in small radii in the thinner grades: heat forming in certain grades to simple shapes such as cylinders and curves: inlaying: engraving: printing by offset process: hot stamping.

General Applications — These may be classified under interior constructional, decoration, furnishing, and illumination. They are used for wall and ceiling panelling, inlaid murals, partitions, doors, telephone booths, lift cages, lift doors, skirting boards, window sills, built-in furniture, counters, counter-tops, bars, bar-tops, and table-tops. The colours and the various surface finishes available, give them the widest possible decorative and practical applications for all sorts of interior work for hotels and restaurants, for shops, transport services and homes. They present a hard, smooth, non-abrasive surface, which is cleaned by simply wiping over with a damp cloth, and this surface may be highly polished, satin-finished, matt, textured or decorated in various ways. The surface texture can be formed by the use of a textured paper as the top layer of a lamination or by using textured plates in the veneer press: either method secures a permanent finish. Patterned papers or textiles can also be used for the top sheet to give a patterned surface, or hand-painted sheets can be used in the same way and after lamination become a permanent and fadeless decoration. Inlaying can be carried out in a great number of colours, and in a range of designs from the most intricate, fine scroll-work to solid blocks of bold colour. Metal foils can also be used in these inlays, and the colours vary from white and pale pastels to all the brighter and darker hues, so that endless variety of effect is possible.

Their resistance to fruit acids and alcohol makes these laminates practical and decorative for bar-tops and restaurant and café tables: there is also, for this particular use, a special heat-resistant grade, which is made by incorporating a thin sheet of aluminium foil next to the top layer of the lamination. If a cigarette or lighted match is left to smoulder on this, no harm is done, for the thin layer of metal diffuses the heat evenly over the entire surface.

The translucent grades are much used for decorative lighting fixtures, for Venetian blind slats, directional signs, illuminated displays and advertising signs. Fluorescent pigments can be added to the translucent laminates for special forms of lighting, and this type is also used in aeroplanes for illuminated dials, instrument panels, nameplates and switches.

See Section III, plates xxix and xxx.

PLASTIC FILAMENTS AND YARNS

Plastic yarns and filaments are of two types: those which are actually extruded from plastic material, and those where a cotton or rayon thread or cord is permanently coated with plastic, retaining full flexibility or being as stiff as wire, as desired. (Strictly speaking, rayon is a plastic filament, but its uses are too familiar to be included in this survey).

NYLON

British Trade Name — Nylon.

American Trade Names — Nylon: Prolon.

Characteristics — Extreme elasticity and tensile strength: toughness: resistant to water and dry cleaning fluids: resistant to mildew and moth: as resistant to light (natural and artificial) as corresponding types of material in silk: can be dyed.

Available Forms — Yarns, bright or de-lustred: monofilaments.

Fabrication Methods — For the yarns, hand- and power-loom weaving; machine and hand knitting: twisting. For the monofilaments, cutting into appropriate lengths for use as bristles.

General Applications — The production of nylon is the result of many years of intensive research to reproduce artificially the molecular chain of real silk. Actually, nylon is not chemically identical with any known natural product, and for this reason should always be thought of as a separate material and not as a synthetic silk. It has different forms and even different species; when woven or knitted into material, it possesses the most delightful softness and warmth of "handle," and gives long and hard wear. It has, so far, been almost exclusively used for the making of stockings and gossamer lightweight lingerie, for the reason that this outlet has taken up all the yarn so far available. But during the war, nylon yarns have been used for making extremely durable and tough material from which parachute covers are made, and this particular type of fabric has great possibilities in the clothing trade, for men's shirts, women's blouses and dresses. The great tensile strength of nylon means that even the very sheerest materials made of it, are hard wearing; for instance, the finest gauge nylon stocking cannot ladder, and will give as long, or longer, hard wear as service weights in other materials. The resistance of the yarn to water is also an asset, for when wet it does not become fragile, and needs no particular care in washing; it dries in an astonishingly short time. The yarn is also made into surgical sutures, fishing lines, sewing thread, bead threading cord, and tennis racquet strings.

H

A certain amount of nylon yarn has been used to produce a lightweight fleece cloth made with a slight admixture of wool. This is extremely warm and soft, resists weather and water, and is easily dry cleaned. It needs daily brushing, and should be steamed instead of pressed when it becomes creased. This material is attractive, both in appearance and "handle," and makes ideal sports topcoats.

The monofilaments are in general use as bristles for all sorts of brushes, both domestic and industrial. The uniformity of diameter of the monofilaments gives the bristle made from it uniform stiffness, and their resistance to chemicals and water is another asset. Laboratory tests show that nylon bristles have double the life of natural bristles.

The clear white, slightly lustrous appearance of the bristles is very attractive. For domestic use, these bristles are widely used for toothbrushes, hair, clothes and nail brushes, and various types of household cleaning brushes. Two examples of their industrial use are: paint brushes (where their resistance to chemicals gives them a long life compared to other bristles), and as inside and exterior bottle-washing brushes. The exterior bottle-washing brush is used in a washing machine for removing labels and dirt from several bottles at one time; the brush is 66 inches long, and bristles successfully withstand the hot caustic solution used, without losing any of their stiffness. Because of their abrasion-resistance they are said to last four times as long as the Tampico fibres which they have replaced.

Another use for nylon bristles is for U.S. Navy gun cleaning brushes, for swabbing out the cores of 16-inch guns. Tests have shown that for this purpose the nylon bristles have a life seven times as long as ordinary bristles before becoming too gummy from the kerosenes or light oils used: when they become clogged, five minutes' boiling in water restores them to full usefulness, a procedure impossible to ordinary bristles. Nylon bristles are also used in the U.S. Navy, for cleaning brushes for 20 millimetre pom-pom anti-aircraft guns, and the .50, .30 and .22 calibre guns for naval aircraft and small ships. (A note on the use of nylon as a moulding powder appears on page 86).

See Section III, plates IV and XXXVII.

POLYVINYLIDENE CHLORIDE

British Trade Names — None.
American Trade Name — Saran.
Characteristics — Resistant to acids, alkalies, salt solutions, organic

solvents and water: resistant to bacteria and fungi: great tensile strength: resistant to fatigue and abrasion: wide colour range.

Available Forms — Extruded monofilaments.

Fabrication Methods — Power- and hand-loom weaving: machine knitting and braiding: twisting.

General Applications — This plastic is one of the two most important which can be used for textiles, the other being nylon. Polyvinylidene chloride filaments are chiefly used for weaving extremely strong cloths for the furnishing trade. The material is lustrous and has a pleasant "handle." Many colours are available. As these are not *dyed* colours, but are actually incorporated in the plastic material from which the filaments are extruded, they have depth and richness and are very fast. Any type of weaving operation can be carried out with the filaments, which do not need to be spun, but are wound ready for use directly on to reels as they come out of the extrusion machine. They can also be used in conjunction with other textile filaments and they are often so used to accentuate some part of the weaving design with their own particular sparkle and lustre. The materials woven from them are available for all sorts of home furnishing uses, and are also extensively used for car, taxi, 'bus, train, ship and aeroplane upholstery. They are light in weight, and yet survive the hardest wearing conditions without losing freshness of appearance.

As the filaments are resistant both to water and chemicals, materials woven from them are easy to keep clean by wiping with a damp cloth. A bottle of ink may be split over such material without doing any harm; again a damp cloth will remove all trace of the accident. All sorts of decorative braids, tapes and cords are made of the filament, also immensely strong and non-corrosive ropes for industrial purposes.

Apart from the furnishing fabrics and industrial ropes, the filaments are also used to make water- and chemical-resistant filter cloths for industrial plants, and for window screens to replace the ordinary wire mesh. These screens are colourful and impervious to atmospheric conditions: they resist corrosion, rust, mildew and fungoid growths. Tropical tents have also been made from this flexible screen material, and these admit air and keep insects out. The inner soles of U.S. Army heavy tropical rubber and other boots are woven from the filaments in a fairly heavy gauge, to keep the foot from contact with the actual sole of the boot. This allows the circulation of air, and the sole can be removed and washed with soap and water.

(A note on the moulding uses of polyvinylidene chloride appears on page 83).
See Section III, plates III, XXXI, XXXII and XXXVI.

POLYVINYL CHLORIDE

British Trade Names — None.
American Trade Name — Vinyon.
Characteristics — Tensile strength: permanently water-resistant: thermo-plastic, *i.e.* a temperature above 65°C. will cause shrinkage: resistant to mineral acids and alkalies: stable to sunlight: can be dyed.
Available Forms — Yarns, bright or de-lustred.
Fabrication Methods — Power- or hand-loom weaving: machine knitting.
General Applications — Industrially, these yarns are used to make dental floss, filter cloths for chemical plants and for screen cloths. They are used to warp-knit shoe uppers for women's wear; gloves made of the yarn are soft and pleasant to wear, non-absorbent, and easy to clean; nor will they stretch. The yarn is also combined with other textile threads in weaving, usually to produce a crease-proof material; for this they are much used in a mixed crease-proof material for men's ties.

CASEIN

British Trade Names — None.
American Trade Name — Aralac.
Characteristics — Some characteristics similar to wool, but not as strong, especially when wet: can be readily blended with other textile threads: can be dyed.
Available Forms — Yarns.
Fabrication Methods — Power-loom weaving.
General Applications — Owing to its lack of strength, this plastic fibre is never used alone, but always in conjunction with other textile threads, chiefly wool and rayon. It is claimed that this admixture produces more economical cloths, as the plastic yarn is considerably cheaper than either wool or rayon, and also that it imparts certain other desirable characteristics. When used with wool, it gives great softness and good draping qualities, and with rayon, unusual types of texture and lustre. It is used considerably in the rayon mixtures for women's dresses; suits made of the wool and casein blend are said to wear well, and to retain their shape,

the cloth being far less expensive than pure wool. It can also be blended with mohair and cotton textile threads with attractive effect. It is also used as an admixture in the production of hat felts. It is used for the strips round which hair is wound before the permanent waving operation. (A note on other uses of casein appears on page 89).

PLASTIC COATED YARNS

British Trade Names — None.

American Trade Name — Plexon Yarns.

Characteristics — Flexible or stiff as required: great tensile strength: water-proof: vermin-proof: rot-proof: weather-proof: flame-proof: resistant to mild acids, perspiration, oil, grease and petrol: will not crack: colour range of 120 shades: opaque, transparent or translucent coatings: rough or smooth surface.

Available Forms — Yarns in thicknesses from .008 inch diameter to .09 inch diameter, in both flexible and stiff qualities: in skeins, hanks, or wound on cones and reels.

Fabrication Methods — Machine and hand weaving: knitting: braiding: crochet and knotting.

General Applications — The method of coating textile threads or cords with plastic originated in France, but has been brought to full and successful commercial development in the last few years in the United States. Cotton, rayon, or glass fibre can all be coated, but rayon is the most used owing to its softness and smooth surface which make it particularly suitable for the process. The coatings can be thin or thick as required, up to 24 coats being possible. The threads are coated by running them through a large composite machine with a trough at each end, which holds the plastic, and a central set of dies and a drying chamber. As the threads come through the plastic-filled trough, they are drawn through the dies, and from here to the drying chamber, and then through the further trough. They can be turned and run through this operation as often as needed to give the number of coats desired, passing through the drying chamber between each coat. Each die is progressively larger, and different shapes can be given to the coated thread according to the shape of the dies; the possible shapes run from an absolutely round thread to sections that are elliptical, triangular or square.

The plastics used are of various kinds, cellulose acetate and cellulose acetate butyrate being those most frequently employed. The stiffness, elongation and yarn diameter are controlled by the number and type of coatings applied.

The coated threads are used to produce all sorts of materials from which belts, braces, handbags, shoe uppers, millinery, dresses, loose covers for furniture, car upholstery and curtains, are made. Lace can also be made from the fine diameter yarn. In the flexible grades, the materials produced by any of the fabrication methods already described are strong, supple, with excellent draping qualities and easily cleaned by wiping over with a damp cloth. The beauty of these coated yarns is remarkable; highly decorative effects can be obtained by combining threads coated with opaque, transparent or translucent plastics. If a coloured thread is used for coating, and a transparent or translucent plastic applied as the coating, unusual rainbow effects are obtainable.

A special type of cotton coated yarn is used as a substitute for steel and copper wire, for making fly screens; this is quite stiff, and can be woven on ordinary wire looms, with very little change-over of process. The screens are rust-proof, will withstand high tropical temperatures, can be any desired colour, and need no painting or lacquering. Coated yarns are also used to make conveyor belts for the food industry, food trays for de-hydrating plants, and sieve meshes for use in the milk, butter and chemical industries. A special stiff grade made by coating tape is used to replace rattan and cane upholstery; it has been found to wear much better, as it never frays, and being already permanently coloured, it requires no painting. A standard serving cord has been made for electrical use on board ship; the fibre coated being non-inflammable glass fibre.

See Section III, plates XXXII, XXXIII, XXXIV and XXXV.

SYNTHETIC RESINS FOR COATINGS, FINISHES AND BONDING AGENTS

The uses for the various synthetic resins represent a highly specialised and complicated subject. It is not possible in this review to do more than briefly to indicate some of the more important resins employed.

PHENOL

Baking Finishes — The method is to spray, dip or otherwise coat the object and to harden the film by baking. This gives a hard, glossy surface which is highly resistant to solvents, acids and chemicals in general, and to oils and grease, and has excellent electrical properties. These finishes are used for inner coatings for tanks in the milk, beer and other food industries, and also for insulating and protecting coils of different types. Armature windings which have been treated with them are rigid and stand up to high rotational speeds. Their great heat-resistance makes them very suitable for use on motors running at abnormally high temperatures. They can also be used as anti-corrosion protection for metals, for coating ceramics and other materials. Examples of these uses are the coating of chemical plant components exposed to corrosive atmospheres, wooden and metal components of textile plants, coating aluminium spools for rayon spinning and coating gas meter components.

Paint and Varnish Manufacture — The paints and varnishes made with phenol have great durability and drying speed. Exterior drying, where they can be used, can be reduced from 24 and 48 hours to 4 and 5 hours. They stand up to weathering and exposure to sunlight without loss of gloss and their resistance to water (including salt water) and humidity, is very great. They are stable to soaps and alkalies, dilute acids, and resist oxidation and industrial fumes. Aluminium paint can also be based on them.

Lacquers — These are used as a protection for metals such as silver, nickel, copper, brass and so forth, and are very hard wearing. They can be either colourless or tinted, and have a good, glossy finish. After stoving, they are odourless, and are not affected by water, alcohol, benzene, acetone, amyl acetate and other solvents of this type. They are also unaffected by oils, dilute acids and sea water. As they resist changes of temperature and humidity, they are of particular value in tropical climates.

Bonding Agents for Plywood — Liquid phenolic resins are used for this purpose and special types are available both for the low and high temperature processes. Plywood impregnated and bonded in this way is immensely strong and will not warp, and is suitable both for exterior and interior use. The resin used is stronger than most woods, and is not affected by weather, water or bacterial growths. Typical applications of the resin-impregnated plywood are for decorative interior panelling, furniture, constructional parts for houses, and constructional units for aircraft, motor-boats and other small craft. The impregnated plywood can be moulded, and this form of fabrication is used for a variety of objects from serving trays to aircraft parts. Another form of phenolic wood adhesive is a thin glue which is used for edge-jointing thin wood veneers in building up wide sheets.

Cements — These are used for bonding moulded phenolic plastic objects or laminated materials, and for cementing these to glass, metal or other materials capable of standing up to the required stoving temperature. Instances of the use of these cements are the bonding of grinding wheel segments, the setting of bristles in all sorts of brushes and the fixing of metal caps to electric light bulbs. (There are notes on the other uses of phenol on pages 66 to 70).

UREA

Wood Cement — This gives a durable joint in constructional woodwork, filling the gap where close contact of two surfaces of wood cannot be made. The joints remain strong even when the cement line is as thick as $\frac{1}{20}\frac{1}{000}$ of an inch. No pressure is needed in its use, the joint being simply held in position till the cement hardens; the cement is water-proof, mould-proof, and sets with great rapidity.

Veneer and Plywood Bonding Agent — This is a very strong bonding agent, and will not discolour the wood; it is porous and allows the moisture and air to penetrate owing to its structure. For this reason any swelling due to damp air or dry air shrinkage does not cause stresses to be set up at the joints of the different layers of the veneer. Besides its use for wood, this bonding agent is used for cementing vulcanised fibre and cardboard, and for sticking felt and other fabrics to wood and cardboard. (There is a note on the moulding uses of urea on page 70).

UREA ALKYD

Stoving Enamels — Though urea is also used alone as a base for baking enamels, it is inclined to break down when subjected to sunlight and weather. It is therefore more generally used combined with alkyd in equal parts; the baking enamels produced from this combination are resistant to surface injury, and to oils, greases, waxes, tars, alkalies, water, humidity, sunlight, moderate heat and wear. The enamels can be produced in an excellent colour range, the whites and paler shades being particularly good. Typical uses are: finishes for hospital, kitchen and bathroom equipment, and for refrigeration apparatus; they are also used on washing-machines. Their good outdoor wearing qualities make them excellent for finishes on cars and other vehicles. They are extensively used for coating articles moulded from plastics, which can only be produced in black or the darker colours.

Other Types of Coatings — These include coating of packaging papers, and anti-gas clothing. Another grade is used for waterproofng raincoat cloth; such cloths do not lose this protective quality with age, nor do they become brittle, soft or sticky.

ALKYD

Apart from its use with urea for the manufacture of stoving enamels, alkyd is also used alone for the same purpose. The enamels thus produced are air-drying, and are for interior and exterior use. Apart from the enamels, alkyd is used to manufacture printing inks, metal primers and finishes, to coat fabrics, to make aircraft lacquers, marine paints and water emulsion paints. Many water emulsion camouflage paints are based on alkyd and some of these can be thinned with petrol as well as water. It is also used in the manufacture of varnishes for boat superstructures, for air-drying finishes for army equipment, and as a coating for cloth for anti-gas clothing.

MELAMINE

Baking Finishes — These can be baked quickly at high temperatures without losing gloss or colour, and have great resistance to electrical arc-ing. They do not easily discolour when exposed to light and heat, but have greater powers of colour retention if combined with alkyd. Production line speed can be considerably accelerated by their use, because of their quick curing time at high temperature baking. They are used as finishes for cars, refrigerators, machinery, metal furniture, washing-machines, hospital, kitchen and bathroom equipment. (A note on the moulding uses of melamine appears on page 72).

CELLULOSE NITRATE

Coatings — These have an excellent surface, and are tough and long-lasting. They are only slightly susceptible to dilute acids and alkalies, but decompose on contact with concentrated forms of either. They have good resistance to petrol and water, medium resistance to animal and mineral oils and are susceptible to salt water or spray, alcohol and vegetable oils. They are highly inflammable, and have a tendency to become hard with age, and some tendency to discolour and become brittle on prolonged exposure to light. In spite of these somewhat negative qualities, they have a large number of successful applications, notably in the manufacture of artificial leathers, which are mostly made by coating various weights of cotton fabric with nitro-cellulose. The artificial leathers are widely used for gold and silver evening slippers and bags, for luggage linings, for handbags and soft luggage. Bookbindings made from these artificial leathers are proof against grease, water, vermin and scuffing, and are washable and highly resistant to wear. Different textures may be given by embossing. A heavy grade of this leather is also used for wall coverings and upholstery material. Other uses for the coating are linings to keep collars from wilting, washable window blinds, coatings for packaging paper which needs toughness, durability and resistance to water, or for labels and fancy-box papers. Both "cellophane" and "glassine" paper are coated with cellulose-nitrate to enhance their appearance and durability.

Lacquers — Used as a finish for brass and other metals. Applications include lacquering of brass and other metal beds, and for the manufacture of bronze and aluminium paints for radiators. The lacquers are also used to coat metal slats, refrigerators, metal furniture, bedsprings and casement window frames. (Other uses of cellulose nitrate are given in a note on page 76).

ACRYLIC

Baking Enamels — White enamels made with this resin will withstand extremely high temperatures without discolouration or becoming brittle, and they are unaffected by chemical fumes.

Coatings — These resist salt water, petrol and vegetable and mineral oils. They are used as solvent finishes for metals, textiles and rubber. As a metal finish, they prevent silver from tarnishing, and will protect the surface of aluminium, copper, chromium, zinc and stainless steel. For textiles, their use gives a permanent stiff finish, increases the brightness of the colours and improves the tensile strength and wearing qualities. (A note on other uses of acrylic resins appears under polymethyl methacrylate, page 78).

POLYVINYL CHLORIDE-VINYL ACETATE

This combination retains the toughness of polyvinyl chloride with the added advantage of the ready solubility of polyvinyl acetate. It is transparent, colourless, tasteless, odourless and non-toxic. Below its softening temperature it is very resistant to water and water vapour, mineral acids, alkalies or salts.

Coatings — These have first-rate adhesive and electrical properties and toughness; they also resist weathering and corrosive atmospheres, and may be coloured to almost any tint, including white. They are much used for lining tins for beer, grape-juice, apple-juice and fats, but are not suitable for use for foods needing heat processing, as the coating is thermo-plastic. Other uses include covers for binoculars and white pigmented types for lining metal barrels and pails. Packaging papers are also coated with them, and can be used as liners for bottle caps and closures for food, medicines and cosmetics. Other packaging uses are for coating aluminium foil for cheese wrapping and milk bottle caps, and as a finish for collapsible metal tubes.

POLYVINYL CHLORIDE

Coatings — Resistant to oils, grease and corrosives, these coatings can be applied to either paper or textiles. Papers coated with it are used for protective packaging purposes. It can be applied to cotton, rayon, wool, felt, asbestos and glass-fibre cloth, and textiles thus treated are used for a number of domestic and industrial applications, such as protective clothing. (A note on the other uses of polyvinyl chloride appears on page 81. *See* also plate xxxi).

VINYL ACETATE

Coatings — Transparent, resilient and strong, this coating is used for waterproofing raincoats, hospital sheeting, gas protective fabrics, mattresses and water-bags. It is also available as a prepared dope which can be vulcanised.

ETHYL CELLULOSE

Coating — Developed in response to war-time shipping needs, ethyl cellulose has proved highly successful in use as a protective packaging material for machine and other metal parts. The plastic is applied as a coating by hot dipping, and dries into a skin-tight coat within a few seconds. It affords complete protection from salt water, corrosion and dirt, and also accelerates assembly of the parts upon delivery, as it can be quickly and easily stripped off with no more elaborate a tool than an ordinary pocket knife. Its use also speeds up the actual packaging of parts, replacing as it does, the laborious methods of coating with heavy grease and then wrapping, or alternatively hand-wrapping, and dipping in hot wax. It is claimed that by its use packaging time for parts treated with it shows reductions as high as 80 per cent over other methods. (A note on the other uses of ethyl cellulose is given on page 77).

SYNTHETIC RUBBERS

Although these are plastics, they are not, in their present stage of development, likely to have the same far-reaching effect upon industrial design that the use of other plastics implies. There are many specialised types, and those named here have the widest applications: Buna.S, Chemigum, Hycar, Neoprene, Thiokol.

The whole field of plastics is growing and developing with great rapidity. Obviously, the synthetic rubbers have a vast period of growth ahead of them; but descriptions of their present characteristics and speculations concerning their future attributes are beyond the scope of this section.

FABRICATION METHODS FOR PLASTICS

COMPRESSION MOULDING

The equipment for this kind of moulding consists of an hydraulic press with a hardened steel mould placed between the platens. The mould is made in two parts which are held in perfect alignment by dowel pins. The moulding compound of powder is placed in an unheated state in a mould which is at moulding temperature. It is then closed by hydraulic pressure. The moulding powder becomes plastic when heated, and under pressure flows into every part of the mould, thus reproducing its shape. Any excess material is forced out between the two halves of the mould where it forms what is known as the "flash." Steam is the most usual form of heat employed, but some moulds are heated by hot water pumped through the press, or sometimes by gas or electricity.

If compression moulding is used for thermo-plastic material, the mould cannot be opened until it is cooled and the moulding inside it set. In the case of thermo-setting plastics, the mould can be opened while still hot. The whole cycle is a short one, its actual length depending on the thickness of section, and no finishing operations are usually required for the completed article beyond the removal of the flash. Metal screws or inserts can be put into the mould at the same time as the moulding compound, and when the cure is complete these will form an integral part of the moulded object.

INJECTION MOULDING

This moulding method is the one most generally used for thermo-plastic materials. The granular powder is fed into a hopper from which it passes into a heating chamber. The correct amount to be admitted is controlled by a special device. The heated powder, which becomes viscous, is then forced under high pressure into a cold mould. The mould is kept closed for a time sufficient to allow the plastic mass to cool and set; it is then opened and the moulded object removed.

TRANSFER MOULDING

This is really a form of injection moulding, and is used for thermo-setting plastics. The moulding compound is subjected to heat and pressure and then forced into a hot mould in which it is cured. The only way in which this process differs from injection moulding is that the mould is maintained at a high temperature once the compound is placed in it, and the moulded parts are ejected from it before cooling.

IMPRESSION MOULDING

This is a fabricating technique used for producing hollow seamless articles in polythene plastic. The mould is preferably in heavy gauge brass and is in two parts, the main tubular body and the cover with a narrow neck. The mould, when assembled, has the appearance of the required article. The inside surface of the mould assembly is painted with a 20 per cent solution of P.84 silicate of soda to prevent adhesion of the polythene to the mould body. At least six times the weight of polythene is required for a given thickness of coating; it is placed in the tubular body of the mould and the whole placed in an oven at 140°C.

When the polythene and mould have reached this temperature, the lid of the mould is placed on the tubular body and the whole assembly is quickly inverted. The molten polythene will slowly travel down the sides of the mould and emerge through the neck, leaving a smooth lining in the mould. During draining the temperature is allowed to fall and cooling is applied to the base. When draining ceases the assembly is immersed in cold water. When cold the lid can be easily removed and the polythene article withdrawn from the mould. The thickness of the wall depends on the rate at which the temperature is allowed to fall and on the viscosity of the polythene.

PRE-FORMING

For this method the moulding compound is shaped into tablets or pills of a definite weight; these pre-forms are placed on a slotted loader plate which is placed over a die or mould. A slider is pushed forward which causes the pre-form to drop into the aligned cavity of the mould. Pre-forms approximate to the shape of the completed moulding. The advantages of this method are that as the pre-form is bulked the press can be closed faster, compression allows the use of a bulk of powder which, as such, would be too great in volume for the loading pot, and as the pre-form shape approximates to that of the mould, there is less need for flow of material, and curing time is therefore shorter.

EXTRUSION MOULDING

There are two different types of this process, called Continuous Dry Extrusion and Continuous Wet Extrusion. In the former, a plastic material either in powder, granular, cube or sheet form, is fed continuously into the extruding machine; production is almost uninterrupted and somewhat resembles squeezing toothpaste out of a tube. The plastic material is spiralled through a heated cylinder by mechanical means, to form a viscous mass. This mass is then forced by pressure through an appropriately shaped orifice in the die which controls the shape of the completed extrusion. That is to say, the extrusion can be in the form of a strip, tube, rod, thread, or profile section according to the shape of the orifice through which the material is forced. As the hot plastic emerges from the die it is air-cooled and carried off on a conveyor belt to winding or cutting machines, according to the form of the extrusion.

In Continuous Wet Extrusion, the initial preparation of the plastic material to be extruded is of the first importance. It is placed in a steel mixer and thoroughly kneaded. It is next filtered to remove any impurities. At this stage it can be put straight into calendering rolls or re-mixed to reclaim the solvents before being calendered. Colouring matter is uniformly absorbed during the rolling or calendering process. Preparation for the actual extrusion is done in two ways, depending on whether a hydraulic or screw type extrusion machine is to be used. For the hydraulic machine, the thick, blanket-like mass into which the plastic has been converted by the calendering rolls, is formed into a long cylinder which is forced into a long cylindrical tube forming part of the extrusion machine. The head of this tube contains the die; this is closed, and steam is applied to the nozzle; when the plastic mass is sufficiently softened by the steam, a large hydraulic ram slowly forces it out through the die at the nozzle, and the finished extrusion, rod, tube or sheet, appears. The mechanical screw machines are also made with a long cylindrical tube; into the back of this are fed chips which have been cut from the blanket-like calendered mass. A screw conveyor carries these chips forward, pushing them through the die orifice. It is possible to extrude two or more distinct colours in separate lines at one and the same time.

CASTING

This is a method much used for certain types of phenolic resin, and as the moulds are of lead or equally inexpensive materials, it allows economical production of short runs of objects, which would be too costly by ordinary moulding methods, where the steel moulds are a heavy initial expense. The casting moulds used are of two types: straight and split moulds. The bulk of casting is done in straight moulds. These are open moulds into which the resin is poured in a syrupy consistency; the moulds are then placed in small ovens, and curing (which may take several days) is done at a medium, controlled temperature. Scallops, flutes and similar designs may be incorporated in the mould, but undercuts are not possible.

The split moulds are made in two pieces which are firmly clamped together; the resin is poured into the mould through a small aperture. Curing then takes place in the usual way. Undercuts, provided that they are not more than two-directional, are possible with this type of mould. An autoclave for fast curing under pressure may be used instead of ovens, with both types of mould. Cored moulds are also used to produce half-spherical castings, pilasters and other objects with a decorated face. The core, which is of metal, is withdrawn only after curing, and before the finished object is removed from the mould.

Phenolic resins are very often cast in bars of special section, which are afterwards sliced, very much as a long loaf of bread is sliced. These slices of the sectional bar are afterwards finished by various machining operations, the slice being a kind of very close "pre-form" of the finished article. Clock-cases, cutlery handles, hair-brush backs, are some of the objects which can be made in this way.

Large, flat open moulds are also used for casting sheets of the glass-like plastics such as polymethyl methacrylate.

FORMING

In this process heat of a sufficiently high temperature is applied to a thermo-plastic sheet, strip, rod or tube to cause it to soften so that it may be bent, twisted or otherwise shaped. The object thus formed is then cooled and retains the new shape given to it.

The forming of large sheets such as those used for various types of aircraft enclosures, is done in the following way: The sheets are hung vertically in an oven until they are soft enough to shape. They are then draped over cloth-covered wooden forms and clamped into position and left to cool on the form. The sheets become rigid and retain the shape given to them by the form. The sheets are also sometimes heated by immersion in baths of hot mineral oil, or by infra-red lamps, where the quantity to be produced does not justify the expense of the ovens.

Strips, rods and tubes are heated on steam-tables, by an electric plate, in small ovens, or by simply plunging in boiling water; they can be given a large variety of shapes by bending, twisting, or other hand operations. The legs of plastic tables and chairs, are for instance, often fastened by binding around them long tubes of the heat-softened plastic, which set when cold and form undetachable fastenings.

DRAWING or SWAGING

This is a form of fabrication used for sheet thermo-plastics. The sheet is pre-heated on a steam table, electric plate, or by plunging in boiling water. It is then placed in an unheated mould, which may be of wood or metal. This is rapidly closed so as to shape the sheet before it loses its pliability. When the sheet has cooled sufficiently to be removed from the mould without distortion it is taken out and thrown into cold water to finish cooling.

DEEP DRAWING

This is a variation of the drawing or swaging method for the thermo-plastic sheet, already described. By means of deep drawing, more intricate shapes can be made than by the ordinary process. The thermo-plastic sheet is pre-heated as for ordinary drawing; it is then stretched into the shape of the model by being pushed by a plunger through an opening which is the same shape and size as the mould. The sheet remains in the mould until it is cool enough to be safely withdrawn without distortion, and the final cooling is done in cold water.

BLOWING

This is another process for forming thermo-plastic sheet, and is chiefly used for cellulose nitrate. Two thin sheets of thermo-plastic material are placed between two pre-heated plates which are made with matching half-cavities. Hydraulic pressure is then applied, and a small nozzle is inserted between the sheets. Hot air or steam is forced between the two sheets through this nozzle, which causes the sheets to blow out and take the shape of the mould. The mould is then chilled (with the blown object still inside it) by dipping in cold water, and the finished article is then removed. This is the method used for manufacturing floating nursery toys, Christmas tree balls, and other lightweight hollow objects.

LAMINATING

Another process for sheet plastics: two thin sheets can be laminated together by being placed between the plates of a press. The press is then closed, and heat applied; after the steam is released the platens are chilled, and the laminated sheet withdrawn. The laminating pressing sheets are usually made of chrome-finished steel.

MACHINING

These operations include all the forms of machining which can be applied to wood or stone, and the technique is the same for plastics as for the other materials. The operations include sawing, drilling, grinding, stamping, carving, turning, polishing, and so forth.

METAL PLATING

This is done by electro-plating, and almost any metal may be used, but silver and copper are those in most common use, as they have the most suitable physical properties. Sometimes only a part of the object is plated, so as to form a contrast between the metal surface and the coloured surface of the plastic.

METAL INLAYS

This can be done in two ways; the metal insert may be put into the mould at the same time as the plastic, the latter material gripping it firmly in position as it cools; or the inlay or insert may be applied after the moulding is complete. This is a more complicated process, and generally used where a *surface* inlay only is wanted, not one which extends throughout the thickness of the plastic.

Inlays inserted into a complete moulding are most often used where it is desired to retain the non-conducting electrical qualities of the plastic, which would be nullified by a metal insert throughout its thickness. The method of applying the inlay is to cut a groove, following the required design, on the surface of the plastic. Into this groove a thin strip of metal, half as deep as the groove, is then laid. On this is then placed the metal inlay which is thus supported by the thin strip of metal first laid in the groove; this lower strip is usually of a harder metal than the actual inlay. Pressure is then applied and the lower strip of metal causes the inlay to spread out in all directions, to penetrate the walls of the groove and thus lock itself into position.

SURFACE DECORATION

This includes engraving, printing, roll leaf hot stamping, embossing, and hand colouring.

Engraving — This is done either by hand or by portable engraving machines which use no heat.

Printing — There are very few limitations to printing on plastic sheeting. Four-colour photogravure, half-tone or direct letterpress printing, can all be used. The rate of printing is slower than on paper, owing to the elasticity of the material, which necessitates constant slight adjustments of the rollers to ensure perfect registration. Quick-drying inks must be used owing to the non-absorbent surface of the plastic sheet. Silk-screen printing can also be carried out on plastics.

Roll Leaf Hot Stamping — For this process a heated die strikes the surface of the plastic through a web of roll leaf paper; the latter carries the pigment or dye which is forced into the indent caused by the impact of the heated die or stamp.

Embossing — This is a form of surface decoration applied to plastic sheets, and the usual steel embossing cylinders or plates are used.

Hand Colouring — This form of surface decoration is mostly used for small articles such as costume jewellery, buttons, and cosmetic containers. It is usually applied at the back of a transparent plastic object, and special non-corrosive water-solvent ink is used for this purpose. Cast phenolic objects sometimes have a design slightly hollowed out on the back of the object, which is afterwards filled in by hand with colouring matter. A transparent pastel-toned object is usually thus decorated, and opaque white is mostly used as colouring matter, so that the effect of the right side of the object is of an opaque, pale, self-coloured pattern, in contrast to the transparent surface.

SECTION III
Plastics in Action

WHEN Alexander Parkes took out his first patent on "celluloid" in 1855, he may have had a vision of the changes his work would ultimately bring to life and industry; but ten years later his confident recitation of the tasks that "Parkesine" could perform proves that he fully understood the potency of his inventions and their future ramifications. The prophetic list he gave to his audience at the Society of Arts on that December afternoon in 1865 comes to life in many of the plates of this section.* Among common objects he mentioned: knife handles, combs, brush backs, shoe soles, walking sticks, umbrella and parasol handles, buttons, brooches and buckles. They are all made of plastics to-day; but nobody who listened to that great Birmingham inventor reading his paper could have imagined that they would be so astonishingly unlike things made from existing substances as, for example, the hairbrush with nylon bristles and a back of polymethyl methacrylate, shown on plate xxxvii of this section or the salad spoon and fork of polystyrene on plate xlvii. He had prefaced his list of common objects by a reference to the "almost unlimited" industrial applications of his new material; a claim that may have seemed extravagant to his hearers, for he only supported it by a few indications of uses, such as spinners' rolls and bosses, embossing rolls, pressing rolls in dyeing and printing works, and mentioned such specialised items as tubes, chemical taps and pipes, battery cells and waterproof fabrics; but plate after plate of this section abundantly justifies the use of those two words: "almost unlimited." True, the objects shown in the plates are not made of Parkesine; but that substance was the great progenitor of present-day plastics, and its inventor would perhaps have been more interested than surprised could he have seen plastics in action to-day.

As it is the purpose of this book to examine plastics in relation to industrial design, this final section has been planned to be complementary to the one preceding it; to form as it were an illustrated commentary on Section II. Therefore the forty-eight plates have been chosen to show not only individual examples of industrial

* See page 15.

design, but the use of plastics for a considerable variety of articles. A few have been selected to demonstrate the capacity of certain methods of fabrication; for example the stretch press die with plastic face and honeycombed wood core on plate v, which shows the scale of castings rendered possible with certain types of cast phenolic resin. The plastics used for the various objects illustrated on each plate are printed in italics, thus: *cellulose acetate*, and they are cross-referenced to the pages in Section II where their properties and characteristics are described.

A cautionary note, in Chapter VI of Section I, may be appropriately repeated in this introduction to the plates. It was suggested that many manufacturers may be tempted to venture a little way into the plastics industry in the future, and attempt to do their own fabricating, in the hope that they may not only solve some of their own production problems, but reduce overhead charges by taking outside orders at nominal rates to keep their moulding plant busy. They will buy their experience dearly. A much better and far less costly investment, would be to retain the services of two or three first-class industrial designers, and conduct some development research work in design, with particular reference to plastics, so that the possibilities of these new materials become imaginatively related to the markets the manufacturer hopes to satisfy.

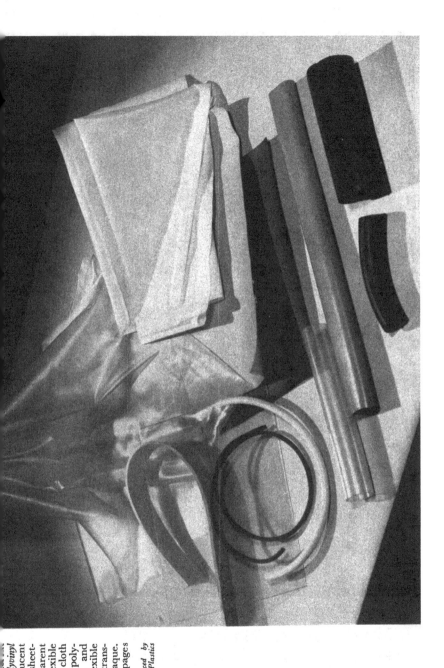

grades of *polyvinyl chloride*: translucent and opaque sheeting, rigid transparent sheet, small flexible strip, cotton cloth coated with polyvinyl chloride, and rigid and flexible tubes, both transparent and opaque. *See* Section II, pages 81 and 103.

Photograph reproduced by courtesy of B. X. Plastics Ltd.

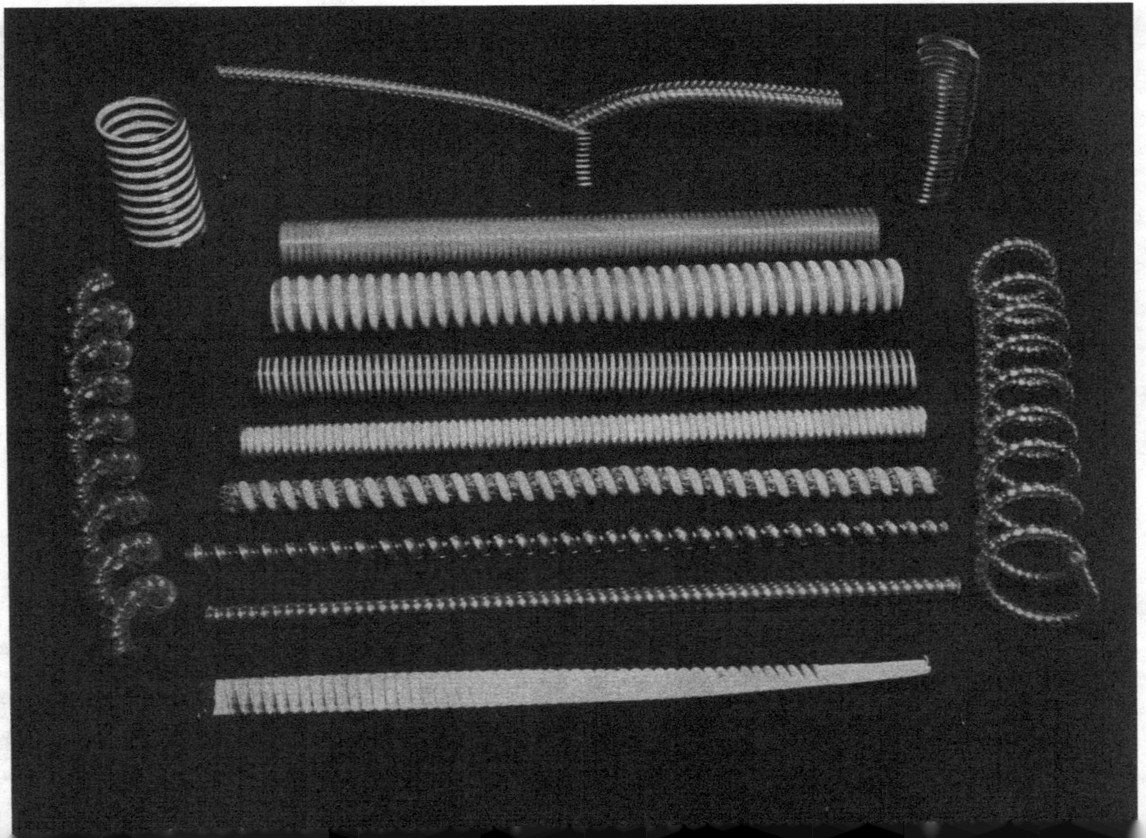

PLATE II Group of tension - wound *cellulose acetate butyrate* extruded strip, and formed into tubes. Some are hollow and used as fixtures for fluorescent lighting; others are mounted on wooden or metal cores for use as table legs, banisters and newel posts. The twisted coils to extreme right and left are made of wire, coated with the same plastic. The flat twisted length in immediate foreground is the cellulose acetate butyrate strip mounted over flat tape for use as lampshade decoration, curtain tie-backs, etc.

(Fabricator: Detroit Macoid Corporation, U.S.A.)

See Section II, page 75.
These examples are photographed by courtesy of the fabricators and Halex Limited.

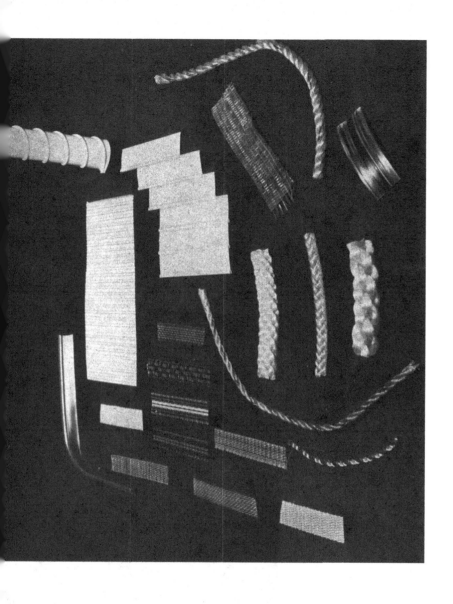

braids at the left are woven from *polyvinylidene chloride* filaments. Above these to the left: extruded flexible *vinyl* strip with upper edge in form of hollow tube (Fabricator: Pierce Plastics Inc., U.S.A.). Top centre: material woven from *cellulose acetate butyrate* strip (Fabricator: Schwab & Frank Inc., U.S.A.), and immediately below this extruded cellulose acetate butyrate strip with interlocking edges (Tennessee Eastman Corporation, U.S.A.). The tube at top right is made of the same extruded strips cemented together. The central cords are made of transparent cellulose acetate butyrate filaments, and extruded strip of the same plastic is used for the small piece of woven material at the right (Fabricator: Schwab & Frank Inc., U.S.A.). The other cords are made of polyvinylidene chloride (Dow Chemical Co., U.S.A.), and the small strip at bottom is a two-colour extruded vinyl strip (Fabricator: Pierce Plastics Inc., U.S.A.).

See Section II, pages 94 and 75.

PLATE IV Above, examples of extruded *cellulose acetate* profile rods showing the variety of section available. The mouldings are used for motor car and furniture trim, linoleum edging, covering wallboard joints, and kindred uses. (Fabricator: Detroit Macoid Co., U.S.A.).
See Section II, page 73.
These examples are photographed by courtesy of the fabricator and Halex Limited.

To the right, toothbrushes with *nylon* bristles; the polymer and uncut nylon bristles are also shown.
See Section II, page 93.
Photograph reproduced by courtesy of I.C.I. (Plastics) Ltd.

PLATE V Stretch press die with plastic face and honeycombed wood core that is visible on the side of the die, used by the Douglas Aircraft Co., Inc., U.S.A. This gives an idea of the size of castings which are possible with certain types of *cast phenolic* resin; castings of this sort weighing nearly 3,000 lbs. have been made. *See* Section II, page 69. *Photograph reproduced by courtesy of Baker Oil Tools Co., U.S.A.*

An assembly jig illustrating the simple methods of manufacturing tools and jigs with *cast phenolics*. It is estimated that 50 to 75 per cent of tooling time and tooling costs can be saved by the use of cast phenolics. *See* Section II, page 69.
Photograph reproduced by courtesy of Baker Oil Tools Co., U.S.A.

PLATE VI To the right, idler gears for water pumps, illustrating the mould in which they are produced and how finished gears are fabricated from the rough casting.
See Section II, page 69.
Photograph reproduced by courtesy of the Catalin Corporation, U.S.A.

Below, thimble sector for a barrage balloon showing how the finished part is cut and fabricated from a special *phenolic casting.*
See Section II, page 69.
Photograph reproduced by courtesy of the Catalin Corporation, U.S.A.

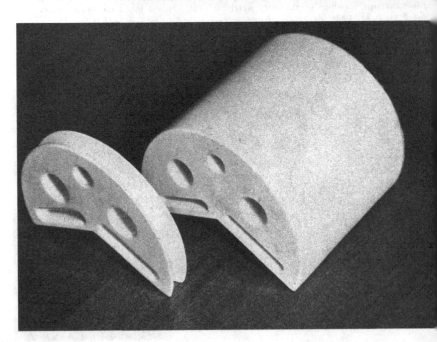

PLATE VII Above, aircraft components machined from *phenolic laminate*.
See Section II, page 90.
Photograph reproduced by courtesy of Bakelite Limited.

To the left, a group of gears machined from *phenolic laminated* silent gear material.
See Section II, page 90.
Photograph reproduced by courtesy of Bakelite Limited.

PLATE VIII Above, high-impact aircraft bolts moulded from *phenol formaldehyde*.
See Section II, page 66. *Photograph reproduced by courtesy of the Monsanto Chemical Company, U.S.A.*

Below, bearing machined from *phenolic laminate*.
See Section II, page 90. *Photograph reproduced by courtesy of Bakelite Limited.*

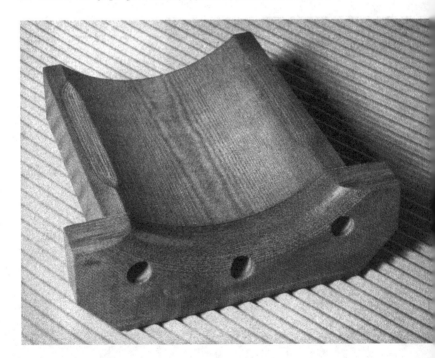

PLATE IX Above, yacht
steering wheel moulded
from *phenolic plastic*.
See Section II, page 66.
*Photograph reproduced by courtesy of
Bakelite Limited.*

To the left, "Vent-axia"
air extractor, showing im-
peller and housing moulded
from *phenolic plastic*.
See Section II, page 66.
*Photograph reproduced by courtesy of
Bakelite Limited.*

PLATE X Above, " Purma"
camera moulded from *phenolic
plastic*.
See Section II, page 66.
*Photograph reproduced by courtesy of
Bakelite Limited.*

To the right, miniature circuit
breaker in moulded *phenolic*
case.
See Section II, page 66.
*Photograph reproduced by courtesy of
Bakelite Limited.*

PLATE XII Above, switch-
ette housing moulded from
nylon.
See Section II, page 86.
Photograph reproduced by courtesy of
E. I. du Pont de Nemours and Company
Inc., U.S.A.

To the right, spool for winding
coils for actuating aircraft
instruments, moulded from
nylon.
See Section II, page 86.
Photograph reproduced by courtesy of
E. I. du Pont de Nemours and Company
Inc., U.S.A.

PLATE XIII Oil spray guard of transparent *cellulose acetate* for automatic screw machines and turret lathes. The oil, instead of spattering every-where, is returned to the drip pans for filtering and re-use. The operator has unobstructed observation of performance. (Designed and fabricated by the Kollsman Instrument Division of Square D Company, Elmshurst, N.Y., U.S.A.).

See Section II, page 73.

Photograph reproduced by courtesy of the Celanese Celluloid Corporation, U.S.A.

PLATE XIV Above, chemical feeder with valve block, pump valve, and re-agent head of *polymethyl methacrylate*. The transparent plastic allows the internal workings of the feeder to be visible, and helps the operator quickly to correct any fouling that may occur.
See Section II, page 78.
Photograph reproduced by courtesy of E. I. du Pont de Nemours and Co. Inc., U.S.A.

To the left, toilet tank float moulded from *cellulose acetate*; it is lighter than a metal float. (Fabricator: Kirkhill Inc., Los Angeles, U.S.A.).
See Section II, page 73.
Photograph reproduced by courtesy of the Celanese Celluloid Corporation, U.S.A.

PLATE XV Lightning arresters of transparent *cellulose acetate butyrate* are used in the U.S. Army Signal Corps communications systems, and other branches of the American armed forces. The housing, which is shatter-proof, protects the electrodes sealed in it from weather, dirt and insects. (Housings moulded by Sterling Plastics Company, Union, N.J., U.S.A., and manufactured by L.S. Branch Manufacturing Corporation, Newark, N.J., U.S.A.).
See Section II, page 75.
Photograph reproduced by courtesy of the Tennessee Eastman Corporation, U.S.A.

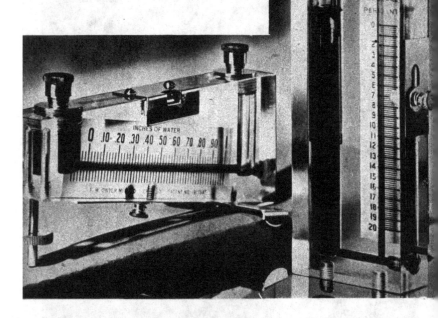

PLATE XVII Above, light-weight *cellulose acetate butyrate* housing for a rotary pneumatic drill. (Fabricator: Reynolds Spring Company. Molded Plastics Division, Cambridge, Ohio, U.S.A., for the Aro Equipment Corporation, Bryan, Ohio, U.S.A.).
See Section II, page 75.
Photograph reproduced by courtesy of the Tennessee Eastman Corporation, U.S.A.

To the left, piano type hinge of *cellulose acetate butyrate*, extruded in continuous lengths. It is lighter than metal, yet strong enough to be used for articles of relatively heavy construction. (Fabricator: Plastics Process Co., Hollywood, Calif., U.S.A.).
See Section II, page 75.
Photograph reproduced by courtesy of the Tennessee Eastman Corporation, U.S.A.

PLATE XVIII   Above, workers' lunch boxes in transparent *cellulose acetate*. ("Transporta" box fabricated by the Irwin Corporation, New York City, U.S.A.).
See Section II, page 73.
Photograph reproduced by courtesy of the Celanese Celluloid Corporation, U.S.A.

Below, transparent, lightweight *cellulose acetate* bait boxes adapted for utility kits and small parts containers. Used for replacement parts shipped abroad, and in factories to concentrate assembly parts and eliminate waste motion. Transparency permits inspection of contents with minimum time and effort, and the boxes are corrosion- and water-proof. (Fabricator: Bill de Witt, Baits Division of Shoe Form Co., Inc., Auburn, U.S.A.).
See Section II, page 73.   *Photograph reproduced by courtesy of the Celanese Celluloid Corporation, U.S.A.*

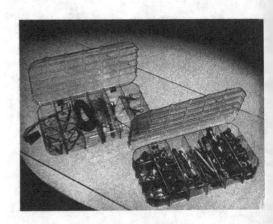

PLATE XIX Ceiling, inset with cast blocks of translucent *polymethyl methacrylate*, lit from behind.
See Section II, page 78.
Photograph reproduced by courtesy of I.C.I. (Plastics) Ltd.

PLATE XX Wall and ceiling lights, and the illuminated backing to the show-cases in this shop, are made from a series of "Lumitile" polystyrene tiles.
See Section II, page 80.

Photograph reproduced by courtesy of Lumitile, Cincinnati, U.S.A.

PLATE XXI Ceiling made entirely of cast *methyl methacrylate* blocks, through which artificial lighting is diffused. (Designed by Timothy Pflueger, San Francisco, U.S.A.). *See* Section II, page 78.

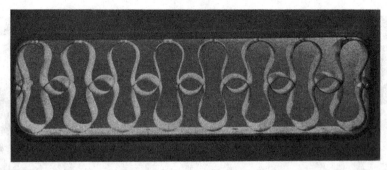

PLATE XXII Above, lighting fixture made of interwoven extruded *cellulose acetate* held in position by cellulose acetate hardware, and set in a frame of extruded cellulose acetate moulding. (Designed by Morris Sanders).
See Section II, page 73.
Photograph reproduced by courtesy of the Celanese Celluloid Corporation, U.S.A.

Below, lighting fixture made of corrugated *cellulose acetate* sheet supported by glass rods held in position by cellulose rings. (Designed by Morris Sanders).
See Section II, page 73.

Photograph reproduced by courtesy of the Celanese Celluloid Corporation, U.S.A.

PLATE XXIII Coach fitted with *polymethyl methacrylate* windows.
See Section II, page 78.
Photograph reproduced by courtesy of I.C.I. (Plastics) Ltd.

PLATE XXIV Above, transparent *cellulose acetate* aircraft components: (1) vision panel (joggled) (2) two deep-drawn light covers. (3) water-tight cemented unit. (Fabricator: Synthena Ltd.).
See Section II, page 73.
Photograph reproduced by courtesy of British Celanese Ltd.

To the right, *polystyrene* gun ports.
See Section II, page 80.
Photograph reproduced by courtesy of the Monsanto Chemical Company, U.S.A.

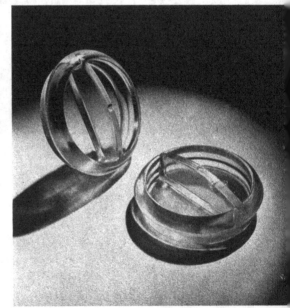

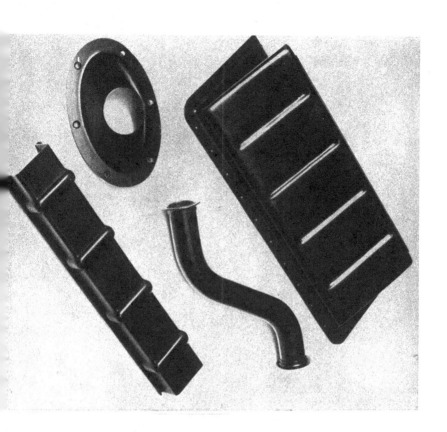

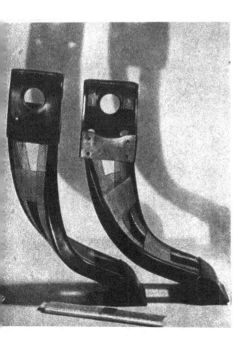

PLATE XXV Above, group of aircraft electrical components, formed from *cellulose acetate* sheet and tube. (Fabricator: Synthena Ltd.).
See Section II, page 73.
Photograph reproduced by courtesy of British Celanese Ltd.

To the left, pair of ammunition feed necks formed from black *cellulose acetate* sheet and wire reinforced *cellulose acetate* sheet. (Fabricator: Thermo-Plastics Ltd.).
See Section II, page 73.
Photograph reproduced by courtesy of British Celanese Ltd.

PLATE XXVI Above, U.S. Army
bugle moulded from *cellulose acetate
butyrate*; it will not dent; the weight
is only ten ounces; no external finish
is applied; and it requires no prac-
tice warming-up notes. (Fabricator:
Elmer E. Mills Corporation, Chicago,
U.S.A., in conjunction with the Chicago
Musical Instrument Co.).
See Section II, page 75.
*Photograph reproduced by courtesy of the Tennessee
Eastman Corporation, U.S.A.*

To the right, U.S. Army canteen moulded
from *ethyl cellulose*. The colour is an
integral part of the material, so no finish
is required. The canteen will endure
extremely hard usage, and the plastic
acts as a thermal insulator.
See Section II, page 77.
*Photograph reproduced by courtesy of the Hercules
Powder Company, U.S.A.*

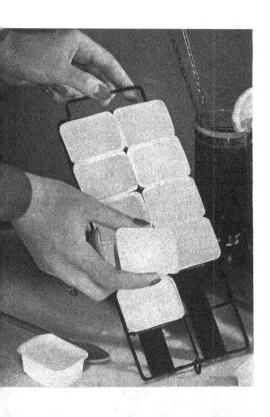

PLATE XXVII To the left, refrigerator ice freezing cups of *ethyl cellulose* sheet. *See* Section II, page 77.

Photograph reproduced by courtesy of the Dow Chemical Company, U.S.A.

Below, ivory coloured *urea* plastic draining board, showing top and reverse. *See* Section II, page 70.

Photograph reproduced by courtesy of British Industrial Plastics Ltd.

PLATE XXVIII
Table crockery used by the U.S. Navy, made from *melamine.*
See Section II, page 72.
Photograph reproduced by courtesy of Modern Plastics, New York City, U.S.A.

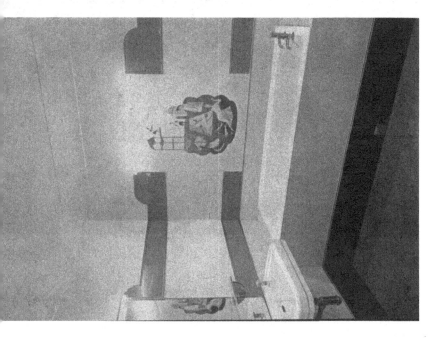

PLATE XXIX To the right, corner of corridor in office suite; partitions are built up of *plastic laminate*, and the door is veneered with *plastic veneer*, incorporating kicking and finger-plates.
See Section II, page 91.
Photograph reproduced by courtesy of Bakelite Limited.

To the left, bathroom walls and bath panels of *plastic laminate*, showing possibilities of inlaid decoration.
See Section II, page 91.
Photograph reproduced by courtesy of Bakelite Limited.

PLATE XXX   Board room in office suite, showing convex panels, skirtings and tiled surround veneered with *plastic laminate.*

See Section II, page 91.
Photograph reproduced by courtesy of Bakelite Limited.

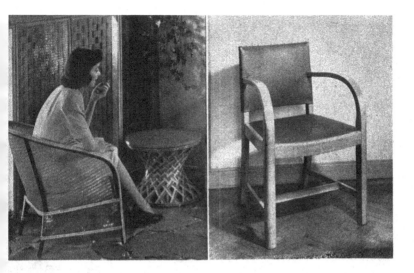

PLATE XXXI *Top right* : Chair upholstered with heavy cotton material coated with *polyvinyl chloride*, in bright emerald green. *See* Section II, pages 81 and 103. *Photograph reproduced by courtesy of B. X. Plastics Ltd.*

Top left : Furniture and window screens of extruded translucent *cellulose acetate butyrate* strip. *See* Section II, page 75. *Photograph reproduced by courtesy of the Detroit Macoid Corporation, U.S.A.*

Top left : Basket-woven material made from extruded *cellulose acetate butyrate* strip for 'bus and train upholstery. (Fabricator: Detroit Macoid Corporation, U.S.A.). *Below this on left* : Power-woven fabric of extruded *cellulose acetate* strip for handbags, millinery and lampshades. (Fabricator: Schwab & Frank Inc., Detroit, U.S.A.). *Lower right* : Furnishing fabric of cotton and rayon with extruded tubes of water white *polyvinylidene chloride* incorporated to give brilliance. (Hand-woven by Mrs. Dorothy Liebes, San Francisco, U.S.A.). *See* Section II, pages 75, 73 and 94. *These examples are photographed by courtesy of the fabricators and Halex Limited.*

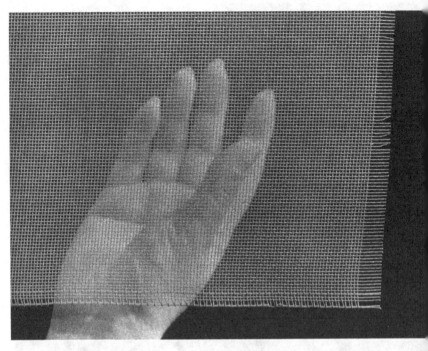

PLATE XXXII Above, insect screen of stiffened cloth, made from "Plexon" *plastic-coated yarn.*
See Section II, page 97.
Photograph reproduced by courtesy of Freyberg Bros. - Strauss Inc., U.S.A.

Below, insect screen, mesh-woven from *polyvinylidene chloride* filament.
See Section II, page 94.
Photograph reproduced by courtesy of the Dow Chemical Company, U.S.A.

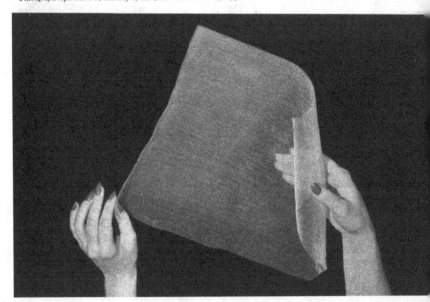

PLATE XXXIII Reels of "Plexon" *plastic-coated yarn*, and a variety of materials made from it. This group illustrates the versatility of the yarn, which can be used on all kinds of weaving machines.
See Section II, page 97.
Photograph reproduced by courtesy of Freydberg Bros. - Strauss Inc., U.S.A.

PLATE XXXIV To the right, handbag of basket-woven *cellulose acetate* strip, and tension-wound bracelet of the same material.
See Section II, page 73.

Below, evening shoes made of "Plexon" *plastic coated yarn*.
See Section II, page 97.

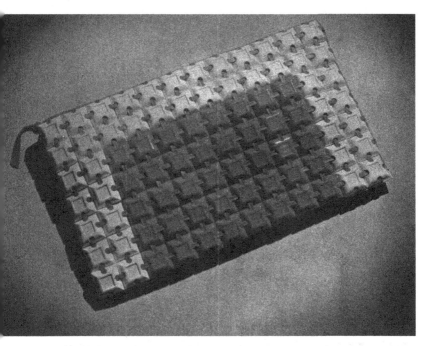

PLATE XXXV Above, handbag made from injection moulded *cellulose acetate* medallions threaded together and mounted on a rayon lining. Such bags in various colour combinations are popular in the U.S.A.
See Section II, page 73. *This example is photographed by courtesy of Halex Limited.*

Below, handbag made of knotted "Plexon" *plastic-coated yarn,* mounted on silk and trimmed with twisted strands of the coated yarn.
See Section II, page 97. *Photograph reproduced by courtesy of Freydberg Bros. - Strauss Inc., U.S.A.*

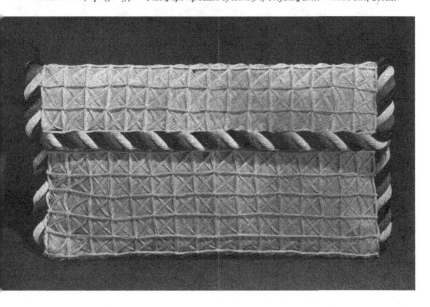

PLATE XXXVI Above, motor car upholstery of woven extruded *polyvinylidene chloride* filaments.

See Section II, page 94. *Photograph reproduced by courtesy of the Dow Chemical Company, U.S.A.*

Below, extremely strong and non-corrosive industrial rope made from *polyvinylidene chloride* strands.

See Section II, page 94. *Photograph reproduced by courtesy of the Dow Chemical Company U.S.A.*

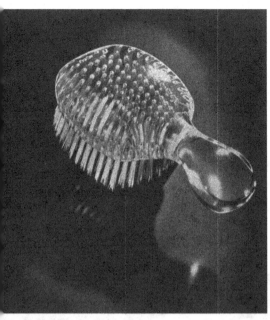

PLATE XXXVII To the left, hairbrush with back of *polymethyl methacrylate*, and *nylon* bristles.
See Section II, pages 78 and 93.
Photograph reproduced by courtesy of E. I. du Pont de Nemours Co. Inc., U.S.A.

Below, brush blocks manufactured from *cast phenolic* resin.
See Section II, page 69.
Photograph reproduced by courtesy of the Catalin Corporation, U.S.A.

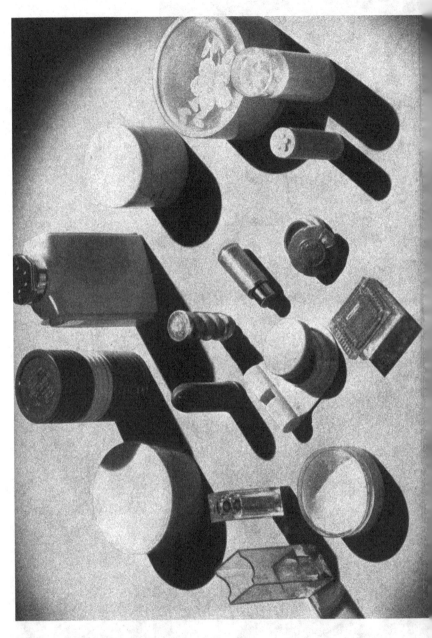

**PLATE
XXXVIII**

American plastic cosmetic containers, including powder boxes, cream jars and perfume bottle and lipstick containers. Materials used are *cellulose acetate, polystyrene, methyl methacrylate* and *cast phenolic*.

See Section II, pages 73, 80, 78 and 69.

These examples are photographed by courtesy of Halex Limited.

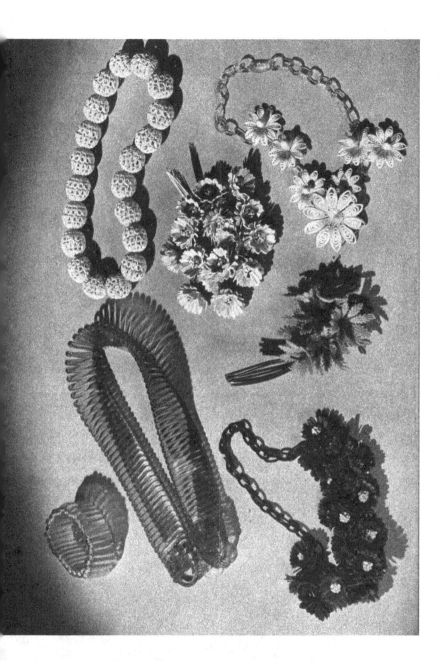

PLATE XXXIX

American costume jewellery at popular prices, made from *cellulose acetate. Top left*: bracelet and necklace made from tension-wound extruded strip, which can be slipped on and off without fastenings. (Fabricator: Schwab & Frank Inc., Detroit, U.S.A.). All other articles are injection moulded and hand assembled, from transparent and opaque plastics in a large range of colours. (Fabricator: The New England Novelty Co., Leominster, Mass., U.S.A.). *See Section II, page 73.*

These examples are photographed by courtesy of the fabricators and Halex Limited.

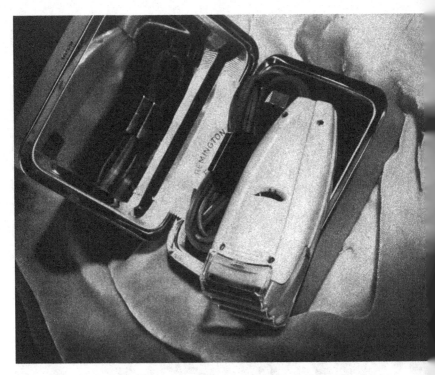

PLATE XL Above, electric razor with *polystyrene* housing; the transparent guard over the cutting head is of the same plastic.
See Section II, page 80. *Photograph reproduced by courtesy of the Dow Chemical Company, U.S.A.*

Below, safety razor housing of *cellulose acetate*. (Fabricator: H. Jamison Company, Freeport, L.I., U.S.A.).
See Section II, page 73. *Photograph reproduced by courtesy of the Celanese Celluloid Corporation, U.S.A.*

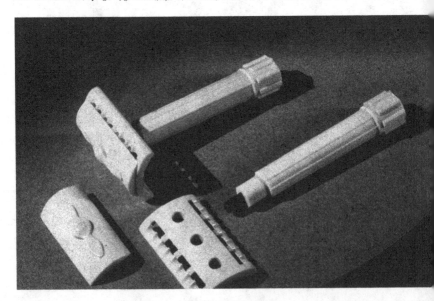

PLATE XLI Above, the two sets of children's building blocks are injection moulded from *cellulose acetate* in bright colours. The other toys, for infants, are sliced from standard *cast phenolic* rods in various opaque colours.
See Section II, pages 73 and 69.
These examples are photographed by courtesy of Halex Limited.

Below, fishing baits moulded from *cellulose acetate*. (Fabricator: The Cruver Manufacturing Company, Chicago, U.S.A.).
See Section II, page 73.
Photograph reproduced by courtesy of the Hercules Powder Company, U.S.A.

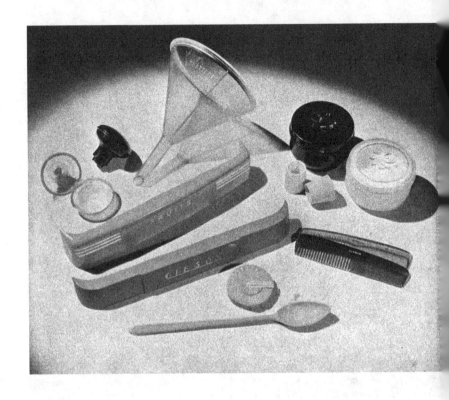

PLATE XLII Articles made from *polystyrene*: door knobs, funnel, cosmetic jars, bottle caps, combs, refrigerator drawer pulls and control dial and spoon.

See Section II, page 80.

Photograph reproduced by courtesy of the Dow Chemical Company, U.S.A.

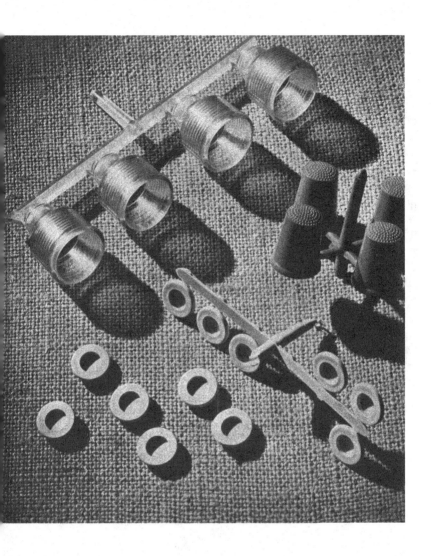

PLATE XLIII Injection moulded *cellulose acetate* tap anti-splashers, glove-buttons and thimbles, showing articles immediately after moulding and before removal of the "stalks".
See Section II, page 73.
Photograph reproduced by courtesy of B. X. Plastics Ltd.

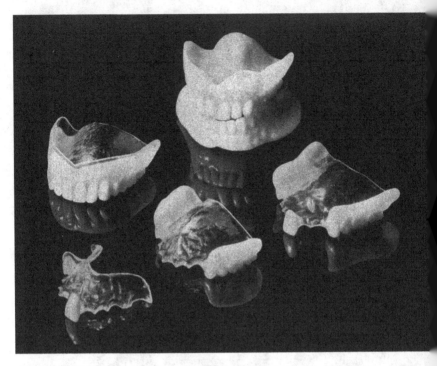

PLATE XLIV Above, dentures made of *polymethyl methacrylate*, showing use of both transparent and opaque types.
See Section II, page 78.   *Photograph reproduced by courtesy of I.C.I. (Plastics) Ltd.*

Below, coat hangers formed from *polymethyl methacrylate* rod and showing technique of permanent fastening by twisting of heat-softened small diameter rods. The curtain hooks are of the same plastic.
See Section II, page 78.   *These examples are photographed by courtesy of Halex Limited.*

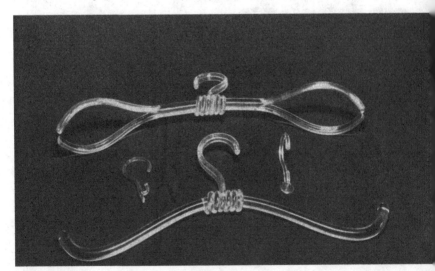

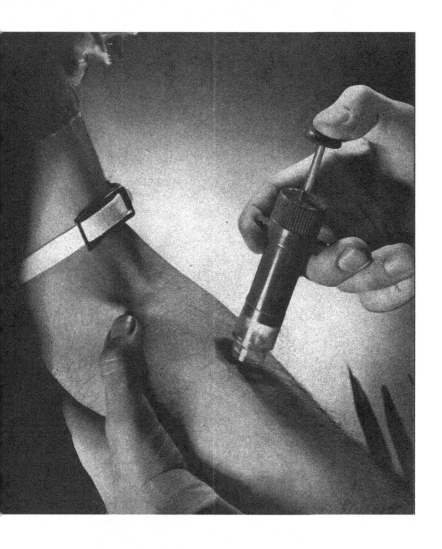

PLATE XLV Vacuum venom extracting pump for snake-bite wounds, moulded from *cellulose acetate butyrate*. Made for U.S. Army for use in tropical warfare, the pump operates under conditions of extreme heat and humidity. (Fabricator: Worcester Moulded Plastics Co., Worcester, Mass., U.S.A., for Sanders Venom Extractor, Tarpon Springs, Fla., U.S.A.).
See Section II, page 75.
Photograph reproduced by courtesy of Tennessee Eastman Corporation, U.S.A.

PLATE XLVI
Transparent containers of *cellulose acetate* sheet.
(Fabricator: Transparent Packings Manufacturing Co. Ltd.).
See Section II, page 73.
Photograph reproduced by courtesy of British Celanese Ltd.

PLATE XLVII

Various articles sold at popular prices in the U.S.A. The hammer head and chisel handle are of amber coloured *cellulose acetate*. Two-colour *vinyl* strip is used for the braces, with white tabs of the same material. The cigarette case is of water white *methyl methacrylate*, moulded with grooves to fit individual cigarettes: the sugar tongs and asparagus tongs are heat-formed from the same plastic. The spoon below these has a two-coloured handle of cellulose acetate, and the salad spoon and fork are of rose-coloured *polystyrene*.

See Section II, pages 73, 78 and 80.

These examples are photographed by courtesy of Halex Limited.

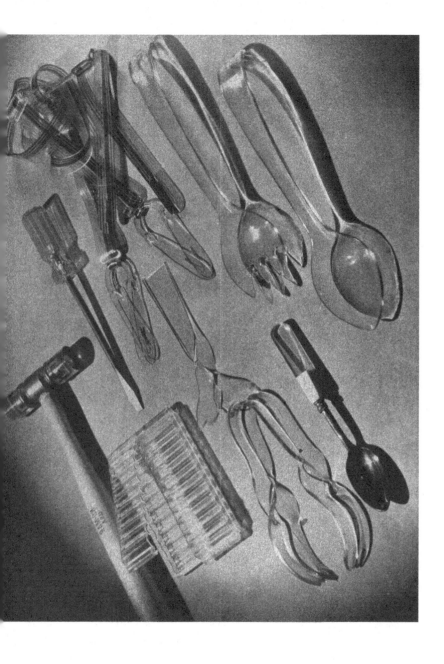

PLATE XLVIII
Radio cabinet made in two colours, of *cast phenolic*.
See Section II, page 69.
Photograph reproduced by courtesy of the Catalin Corporation U.S.A.

Glossary

'A' STAGE RESINS.—First re-action stage of thermo-setting resins, when they are still soluble and fusible; the stage at which they are used for impregnation.

ACRYLIC.—Generic name for plastics made from acrylic acid or its derivatives.

ALKYD.—Resin with a coal tar, napthalene and benzene base. (*See* page 101).

AMINO.—Used as a combining term in the names of chemical compounds, it indicates the presence of chemicals derived from ammonia.

AMINOPLAST.—The term used to describe resins made from amino compounds.

'B' STAGE RESINS.—Intermediary stage of thermo-setting resins at which they soften when heated, and swell in contact with liquids, but do not entirely fuse or dissolve. Thermo-setting powders are in this stage before moulding.

BAGASSE.—Fibrous by-product of sugar cane.

BLENDING.—The mixing of all the ingredients of a moulding compound by mechanical means.

BLOWING.—Fabrication method used for thermo-plastic sheet or film. (*See* page 108).

BULK FACTOR.—Ratio of volume between loose moulding powder and the finished moulded article.

'C' STAGE RESINS.—Infusible and insoluble final stage of thermo-setting resins.

CASEIN.—Protein plastic derived from milk base. (*See* page 89).

CAST.—Fabrication method for plastic material, consisting of pouring it into a mould and curing by heat without pressure. (*See* page 107).

CASTING.—Finished object made by casting method.

CATALYST.—Substance that, without undergoing any change itself, initiates or accelerates a chemical action.

CELLULOSE.—Substance which forms the solid structure of plants.

CELLULOSE ACETATE.—Plastics made from a combination of acetic acid, anhydride and cotton linters. Also used to describe a rayon yarn or fabric with the same base. (*See* pages 73 and 97).

CELLULOSE ACETATE BUTYRATE.—Plastics made from acetic and butyric acids, and cotton linters. (*See* page 75).

CELLULOSE NITRATE.—Plastics made from nitric and sulphuric acids and cotton linters. (*See* page 76).

COLD MOULDING.—Fabrication method in which a plastic compound is shaped at room temperature and cured by baking.

COMPRESSION MOULDING.—Permanent shaping of plastic objects in moulds by the application of heat and pressure. (*See* page 104).

CURING.—Heat setting of a resinoid.

CYCLE.—One complete operation of a moulding press.

DEEP DRAWING.—Fabrication method for plastic sheet by means of forming die and plunger. (*See* page 108).

DRAWING or SWAGING.—Stretching a plastic sheet to reduce thickness; also to give a simple shape. (*See* page 108).

ETHYL CELLULOSE.—Plastics made by treating cotton linters or wood pulp with sodium hydroxide and ethyl chloride or sulphate. (*See* page 77).

EXPANDED PLASTICS.—Plastics with independent or non-intercommunicating cellular construction. (*See* page 87).

EXTRUSION MOULDING.—The manufacture of rods, tubes, strip or profile rods, by forcing a heat-softened plastic moulding compound through a shaped orifice. (*See* page 106).

FILLER.—Various inert materials which are added to a plastic to give it different characteristics such as toughness, impact strength, opacity, etc.

FLASH.—Excess moulding compound forced out from a mould when it is closed.

FORMING.—Application of force to a heat-softened plastic to give it a desired shape. The operation may consist of simple bending or twisting, or of shaping over a wooden mould. (*See* page 107).

IMPRESSION MOULDING.—Fabrication method used for producing hollow, seamless articles with polythene. (*See* page 105).

INJECTION MOULDING.—Fabrication of plastic objects by forcing the heat-softened plastic powder into a cold mould of the desired shape. (*See* page 104).

LAMINATED SHEET.—Two or more sheets of thermo-plastic material bonded together by heat and pressure.

MELAMINE.—Plastic or resin with a base of lime, coke, calcium carbide and nitrogen. (*See* page 72).

MONOMER.—Simplest repeating structural unit of a polymer.

NYLON.—Synthetic linear super-polymer monofilaments, yarn and plastic. (*See* pages 86 and 93).

PHENOLIC.—Generic name covering any plastic of which phenol is the chief ingredient. (*See* page 66).

PLASTIC LAMINATES.—Laminated sheet material impregnated and bonded with thermo-setting resins. (*See* page 90).

PLASTICIZE.—The softening of a material to make it plastic or mouldable.

PLASTICIZER.—Chemical substance added to a plastic compound to make it softer and more flexible.

POLYMER.—Chemical compound of high molecular weight, formed by the combination of simpler compounds, having the same chemical elements in the same proportions.

POLYMERISATION.—Chemical change which produces a new compound, the molecular weight of which is a multiple of the original substance.

POLYMETHYL METHACRYLATE.—Transparent plastic made from acrylic acid and derivatives. (*See* page 78).

POLYSTYRENE.—*See* styrene.

POLYTHENE.—Straight-chain hydrocarbon plastic. (*See* page 85).

POLYVINYLIDENE CHLORIDE.—Plastics with a petroleum, brine and chloride base. (*See* page 83).

PRE-FORMING.—Fabricating process which uses compressed tablets of plastic powder and generally employed for greater speed and accuracy in compression moulding. (*See* page 105).

RESINOID.—A synthetic chemical compound, produced by condensation and polymerisation to form a resinous substance which possesses plastic properties.

SARAN.—Generic name for vinylidene chloride plastic. (*See* page 84).

STYRENE or POLYSTYRENE.—Plastics with a coal and petroleum base. (*See* page 80).

SWAGING.—*See* drawing.

SYNTHETIC RESIN.—An organic solid or semi-solid material built up by chemical reaction and possessing plastic properties.

TREACLE STAGE RESIN.—Liquid state of a casting thermo-setting resin.

THERMO-PLASTIC.—A plastic resin which softens at a given temperature. All plastics soften at the application of heat during the initial stages of their manufacture; thermo-plastics retain this characteristic even after fabrication. (*See* page 65).

THERMO-SETTING.—A plastic which undergoes a chemical change, resulting in permanent hardening, at a certain temperature. Alterations of shape can thereafter only be effected by cutting, sawing, drilling, etc. (*See* page 65).

TRANSFER MOULDING.—A variant of injection moulding, used for thermo-setting plastics. (*See* page 105).

UREA.—General term for plastics with an ammonia, carbon dioxide and hydrogen base. (*See* page 70).

VINYL.—Plastics with a vinyl acetate or vinyl chloride base.

Index

Printed in the United States
by Baker & Taylor Publisher Services